PROGRESS IN
THEORETICAL
PHYSICS

Tenth Annual Montreal-Rochester-Syracuse-Toronto
Meeting on High Energy Theory

PROGRESS IN THEORETICAL PHYSICS

Toronto, Canada
May 9 – 10, 1988

Editors
T Barnes
B Holdom
PJ O'Donnell

World Scientific
Singapore • New Jersey • London • Hong Kong

Published by

World Scientific Publishing Co. Pte. Ltd.
P O Box 128, Farrer Road, Singapore 9128

USA office: World Scientific Publishing Co., Inc.
687 Hartwell Street, Teaneck, NJ 07666, USA

UK office: World Scientific Publishing Co. Pte. Ltd.
P O Box 379, London N12 7JS, England

PROGRESS IN THEORETICAL PHYSICS

ISBN 9971-50-775-7

Printed in Singapore by JBW Printers & Binders Pte. Ltd.

Preface

The tenth annual Montreal-Rochester-Syracuse-Toronto meeting on High Energy Theory was held at the University of Toronto on 9–10 May 1988. These proceedings are a compilation of the written versions of the talks presented at the meeting, and represent contributions to the areas of quantum chromodynamics, electroweak physics, model quantum field theories, string theory, gravity and mathematical physics.

We would like to take the opportunity to express our gratitude to all who assisted with the meeting, in particular the theoretical physics secretaries, Anna Reale and Aphrodite Roussos-Tsironikos.

We gratefully acknowledge the financial assistance of the Physics Department of the University of Toronto.

The Editors

CONTENTS

TESTING THE COLEMAN-HILL THEOREM IN A THEORY OF VECTOR BOSONS COUPLED TO PHOTONS

C.R. Hagen, P.K. Panigrahi
and
S. Ramaswamy

ABSTRACT

In light of the Coleman-Hill Theorem which asserts that the topological mass term in QED_3 gets one-loop correction only from the spinors, we analyze a theory of vector mesons coupled to photons. It is shown that in presence of a parity violating coupling, the topological mass does get a contribution from the vector bosons as well. Although the result is generally divergent, there exists the possibility of obtaining a finite result when a certain sum rule on inverse masses is satisfied. Taking into account the sum rule, a single expression is seen to unify the one-loop contributions from spin-zero, spin one-half and spin one particles to the topological mass.

In $2+1$ dimensional Quantum Electrodynamics, there exists the possibility of adding the so called Chern-Simons (C-S) term to the action. Under a gauge transformation the C-S term changes by a total divergence and therefore can be a legitimate part of the action. In presence of this term photon acquires a gauge invariant mass and hence the C-S QED has attracted considerable attention in the recent literature.[1)] Although the Abelian C-S term, unlike its non-Abelian counterpart does not have any topological significance, the photon's mass so generated goes by the name of topological mass. In interacting theories the radiative corrections to this mass has been studied by many authors.[1,2)] The two novel features that emerged from the perturbative analysis are the following:

i) there is a finite one-loop contribution to the topological mass from the fermions, the two-loop contribution being indentically zero;

ii) there are no contributions from the charged scalars and vector bosons coupled to photons in the conventional way.

In a succinct paper, Coleman and Hill[3)] then systematically studied the general theory of a photon interacting with scalar, spinor and/or vector fields and asserted that to all orders in perturbation theory, all corrections to the topological mass term (beyond the one-loop fermionic corrections) vanishes identically. The proof does not require renormalizability of the theory. The C-S term being parity violating, the lack of contribution from the scalar fields can be ascribed to the absence of parity odd terms in the lagrangian. The fermion mass term in $2+1$ dimensions violate parity and hence is responsible for the radiatively generated C-S term as can be easily seen by a one-loop calculation. In light of the above discussions, it is not clear why an arbitrary parity violating term should not affect the one-loop result. Vector fields in particular offer the possibility of incorporating such parity odd terms in the action and hence such theories need further scrutiny. A careful analysis of ref.3 reveals that such terms have not been included in the vector meson sector of the model.

Here we will study a vector meson theory with parity odd terms in the lagrangian and will demonstrate that C-S terms will be induced at the one-loop level. Two different formulations of the same theory will be contrasted to show the simplicity and elegance of a certain approach over the conventional one.

The pure C-S QED is characterized by the lagrangian

$$L = \tfrac{1}{4} F_{\mu\nu} F^{\mu\nu} - \tfrac{1}{2} F_{\mu\nu}\left[\partial^\mu A^\nu - \partial^\nu A^\mu\right] + \tfrac{\mu}{4}\epsilon_{\mu\nu\rho} A^\mu F^{\nu\rho}. \quad (1)$$

The equations of motion after some manipulations become

$$\Box F^*_\beta + \mu^2 F^*_\beta = 0\,, \quad (2)$$

revealing the massive nature of the excitations.
Here

$$F^*_\mu = \tfrac{1}{2}\epsilon_{\mu\alpha\beta} F^{\alpha\beta}\,, \quad (3)$$

$\epsilon_{\mu\alpha\beta}$ being the Levi-Civita tensor. The general result for propagator of Abelian gauge field coupled to a conserved vector current can be written as

$$D^{\mu\nu}(q) = D_0^{\mu\nu}(q) + e^2 D_0^{\mu\alpha}(q)\,\Pi_{\alpha\beta}(q)\, D^{\beta\nu}(q)\,, \quad (4)$$

the zero subscript denoting the $e = 0$ result and $\pi_{\alpha\beta}(q)$ being the current correlation function. In $2+1$ space-time dimensions gauge and Lorentz invariance allows us to write

$$\Pi^{\mu\nu}(q) = (q^\mu q^\nu - q^2 g^{\mu\nu})\,\Pi(q^2) - i\,\epsilon^{\mu\nu\alpha} q_\alpha\,\overline{\Pi}(q^2). \quad (5)$$

The precise statement of the Coleman-Hill theorem is then

$$\overline{\Pi}(0) = \tfrac{1}{4\pi}\sum_{\text{spinors}} (Q_a)^2 \frac{m_a}{|m_a|}\,, \quad (6)$$

where Q_a and m_a are the charge and mass of the spinors respectively. Hence the principal aim of this work will be the study of $\bar{\pi}(o)$ in a vector boson theory.

In the usual first order formulation, the lagrangian for the vector meson field can be written as

$$L = \tfrac{1}{4} G_{\mu\nu} G^{\mu\nu} - \tfrac{1}{2} G_{\mu\nu} (\partial^\mu B^\nu - \partial^\nu B^\mu) - \tfrac{1}{2} \mu^2 B_\mu B^\mu + \tfrac{1}{2} \lambda \epsilon^{\mu\nu\alpha} B_\mu \partial_\nu B_\alpha , \tag{7}$$

the last term being the parity odd one. Standard analysis reveals the presence of two propagating modes having masses

$$\mu_\pm^2 = \mu^2 + \lambda^2/2 \pm (\mu^2 \lambda^2 + \lambda^4/4)^{1/2} . \tag{8}$$

The propagator can thus be written as

$$G_{\mu\nu} = \sum_{i=\pm} G^i_{\mu\nu} , \tag{9}$$

where

$$G^{\mu\nu}_\pm = \frac{\pm}{p^2 + \mu_\pm^2} \left(g^{\mu\nu} + \frac{p^\mu p^\nu}{\mu_\pm^2} + i\lambda \frac{\epsilon^{\mu\nu\alpha} p_\alpha}{\delta\mu_\pm^2} \right) \frac{\delta\mu_\pm^2}{\delta\mu^2} \tag{10}$$

and

$$\delta\mu_\pm^2 = \mu_\pm^2 - \mu^2 , \quad \delta\mu^2 = \mu_+^2 - \mu_-^2 . \tag{11}$$

In principle it should now be straightforward to calculate $\pi(0)$ to one-loop order, when the vector meson is coupled minimally to the electromagnetic field. This however turns out to be an extremely tedious task, which can be seen by inspecting the structure of the current correlation function $\pi_{\mu\nu}(q)$ given below.

$$\begin{aligned} \Pi^{\mu\nu}(q) = i \int \frac{d^3p}{(2\pi)^3} \Big\{ & \big[2ip^\mu g_{\alpha\beta} - i\delta^\mu_\alpha (p+q/2)_\beta - i\delta^\mu_\beta (p-q/2)_\alpha \\ & - \lambda \epsilon_{\beta\rho\alpha} g^{\rho\mu} \big] G^{\alpha\beta}(p+q/2) \big[2ip^\nu g_{\sigma\tau} - i\delta^\nu_\tau (p-q/2)_\sigma \\ & - i\delta^\nu_\sigma (p+q/2)_\tau - \lambda \epsilon_{\sigma\rho\tau} g^{\rho\nu} \big] G^{\tau\beta}(p-q/2) + 2ig^{\mu\nu} G^\alpha{}_\alpha(p) \\ & - 2iG^{\mu\nu}(p) - \big[1 + \tfrac{1}{24} (q\,\partial/\partial p)^2 \big] \big[\partial/\partial p_\nu (2ip^\mu g_{\alpha\beta} - i\delta^\mu_\alpha p_\beta \\ & - i\delta^\mu_\beta p_\alpha - \lambda \epsilon_{\beta\rho\alpha} g^{\rho\mu}) G^{\alpha\beta} \big] + i g^{\mu\nu} \tfrac{1}{4} (q\,\partial/\partial p)^2 G^\alpha{}_\alpha(p) \\ & - i/8\, (q\,\partial/\partial p)^2 \big[G^{\mu\nu}(p) + G^{\nu\mu}(p) \big] \Big\} . \end{aligned} \tag{12}$$

It can be checked by a straightforward calculation that

$$\Pi_{\mu\nu}(q) = \Pi_{\nu\mu}(-q)$$

and

$$\Pi_{\mu\nu}(q)\, q^{\nu} = 0 \qquad (13)$$

Apart from the enormous amount of algebraic effort to carry out the computation (which the authors have done) the calculation is plagued with spurious divergences and the associated need to develop a consistent set of rules for the translation of integration variables.

In light of the above facts, it is pertinent to find out what is the most general way to minimally couple the massive vector bosons to electromagnetic field. Following an earlier approach applied to the case of gauge fields[5)] and noting that $G^{\mu\nu}$ can be written as the dual of a vector in $2+1$ dimensions, the most general form of the lagrangian can be written as

$$L = \tfrac{1}{2}\psi^{\mu} A\, \epsilon_{\mu\nu\alpha}\partial^{\nu}\psi^{\alpha} + \tfrac{1}{2}\psi^{\mu} B \psi_{\mu} \,, \qquad (14)$$

where A and B are real symmetric matrices and $\psi^{\mu} = \binom{\phi^{\mu}}{\Phi^{\mu}}$. Here ϕ^{μ} and Φ^{μ} are arbitrary linear combinations of B^{μ} and the dual of $G^{\mu\nu}$. The doublet field ψ^{μ} is taken to be Hermitian. B can be brought to a diagonal form with nonzero and nonpositive eigenvalues by an orthogonal transformation of ψ^{μ}. The signs of the eigenvalues are dictated by the requirement of positivity of the energy density. Scaling of the eigenfields brings B to a multiple of the unit matrix, allowing us to do another orthogonal transformation to diagonalize A while keeping B invariant. Since the eigenvalues of A can be of either sign (the zero eigenvalue will not be considered as it will play no dynamic role) a subsequent rescaling of fields allows us to write

$$L = -\tfrac{1}{2}\mu_1 \phi^{\mu}\phi_{\mu} - \tfrac{1}{2}\mu_2 \Phi_{\mu}\Phi^{\mu} - \tfrac{1}{2}(-1)^{s(\phi)}\phi^{\mu}\epsilon_{\mu\nu\alpha}\partial^{\nu}\phi^{\alpha} - \tfrac{1}{2}(-1)^{s(\Phi)}\Phi^{\mu}\epsilon_{\mu\nu\alpha}\partial^{\nu}\Phi^{\alpha} \,. \qquad (15)$$

Here $\mu_i > 0$ and $s(\phi)$ and $s(\Phi)$ can independently be either one or two.

The equations of motion are

$$\mu_1 \phi_\mu + (-1)^{s(\phi)} \epsilon_{\mu\nu\alpha} \partial^\nu \phi^\alpha = 0 ,$$

$$\mu_2 \Phi_\mu + (-1)^{s(\Phi)} \epsilon_{\mu\nu\alpha} \partial^\nu \Phi^\alpha = 0 . \qquad (16)$$

As can be readily seen when the signatures of the two fields satisfy $(-1)^{s(\phi)+s(\Phi)} = 1$, the eigenfields ϕ^μ and Φ^μ have same polarizations and the polarizations are opposite for the opposite sign. The usual parity conserving theory can be recovered by choosing the signs in front of the kinetic terms for ϕ^μ and Φ^μ fields to be positive and negative respectively and carrying out the following steps. The masses μ_1 and μ_2 are chosen to be same and the fields are redefined as

$$(\phi_\mu - \Phi_\mu) = B_\mu \quad and \quad (\phi_\mu + \Phi_\mu) = \epsilon_{\mu\nu\alpha} G^{\nu\alpha} . \qquad (17)$$

The vector fields can now be coupled to the gauge field A_μ by minimal coupling and each field will make separate contributions to $\pi(0)$. The Green's function for the polarization s is

$$G_{\mu\nu}(q) = \left[\mu_s g_{\mu\nu} + 1/\mu_s \, q_\mu q_\nu + i(-1)^s \epsilon_{\mu\nu\alpha} q^\alpha \right] \frac{1}{q^2 + \mu_s^2} . \qquad (18)$$

Computation of $\pi_{\mu\nu}(q)$ is now straightforward and is given by

$$\pi^{\mu\nu}(q) = i \int \frac{d^3p}{(2\pi)^3} \epsilon^{\beta\mu\alpha} \left\{ G_{\alpha\sigma}(p + q/2) \epsilon^{\sigma\nu\tau} G_{\tau\beta}(p - q/2) - G_{\alpha\sigma}(p) \epsilon^{\sigma\nu\tau} G_{\tau\beta}(p) \right\} . \qquad (19)$$

It is worth mentioning that the lowest order vertex function is momentum independent and is given by the Levi-Civita tensor. This simple form of the vertex leads to an enormous reduction in algebraic effort to compute $\pi_{\mu\nu}(q)$. As can be easily seen the $\pi_{\mu\nu}(q)$ statisfies

$$\pi^{\mu\nu}(q) = \pi^{\nu\mu}(-q)$$

and

$$q_\mu \pi^{\mu\nu}(q) = 0 . \qquad (20)$$

Since our interest lies in the axial vector part of the current correlation function we compute

$$\tfrac{1}{2}q^2\, \epsilon_{\mu\nu\alpha}\, q^\alpha\, \pi^{\mu\nu}(q)\Big|_{q^2=0} = (-1)^s \int \frac{d^3p}{(2\pi)^3} \frac{1}{(p^2+\mu_s^2)^2}\left[\mu_s - p^2/\mu_s\right] \tag{21}$$

Hence

$$\bar{\pi}(0) = i(-1)^s \int \frac{d^3p}{(2\pi)^3} \frac{1}{(p^2+\mu_s^2)^2}\left[\mu_s - p^2/\mu_s\right] \tag{22}$$

The momentum integral is linearly divergent and there are no other spurious divergences as encountered in the earlier formulation. Since the opposite polarizations yield contributions of opposite sign it is indeed possible to cancel the linear divergence by demanding an appropriate condition on the inverse masses. Generalization from the two fields ϕ_μ and Φ_μ to N fields is trivial and in that case the necessary condition is given by

$$\sum_{i=1}^{N} (-1)^{s_i} \frac{1}{\mu_i} = 0 \quad . \tag{23}$$

In this case there is no divergence in the final result and the result simplifies to

$$\bar{\pi}(0) = -\tfrac{1}{2\pi} \sum_{i=1}^{N} (-1)^{s_i} \quad . \tag{24}$$

The net contribution to $\bar{\pi}(o)$ is seen to be proportional to the difference between the number of positive and negative polarizations. It is extremely rewarding to see that the contributions from the scalar and spinor sectors can now be combined with the vector field contribution to yield

$$\bar{\pi}(0) = -\tfrac{1}{2\pi} \sum_{S=0,\frac{1}{2},1} \sum_i S\,(-1)^{s_i} \quad . \tag{25}$$

The fermion charges have been taken to be integral and s_i in this case also takes on two values according to the sign of the mass term. It is remarkable that a single expression unifies the contributions from all the sectors, at the same time allowing for the inclusions of an arbitrary number of fields. Generalization to higher spins is an open questions and is pursued at the moment.

ACKNOWLEDGEMENTS

This work is supported by the U.S. Department of Energy Contract No. DE-AC02- 76ER13065. The authors are particularly thankful to Nadine Catterall for typing the manuscript.

REFERENCES

1. W. Siegel, Nucl. Phys. B156, 135 (1979); S. Deser, R. Jackiw and S. Templeton, Ann. Phys. (NY) 140, 372 (1982).
2. Y. Kao and M. Suzuki, Phys. Rev. D31, 2137 (1985).
3. S. Coleman and B. Hill, Phys. Lett. 159B, 184 (1985).
4. C. R. Hagen, Phys. Rev. D36, 3773 (1987).
5. C. R. Hagen, P. Panigrahi and S. Ramaswamy, University of Rochester Preprint, to appear in Phys. Rev. Lett.

GLUON INTERACTIONS AND DIFFRACTION

B. Margolis, P. Valin
Dept. of Physics, McGill University
3600 University St., Montreal H3A 2T8, Canada

and

F. Halzen
Dept. of Physics, University of Wisconsin
Madison, Wisconsin, 53706, USA

ABSTRACT

We examine the relationship between gluon-gluon scattering and various inelastic *pp* and $\bar{p}p$ processes as well as their shadow, the elastic channel. At $\sqrt{s}$ values in the few hundred GeV region typical hadron collisions become semi-hard and calculable using perturbation theory. An effective threshold develops as a result of the rapidly increasing gluon content of the proton. This results in an increasing total cross section and a relatively large $\rho(0)$, consistent with a recent UA4 measurement.

INTRODUCTION

For some years now we have been involved in the development of a picture of how the parton model dictates the energy dependence and other features of hadron-hadron scattering and total cross sections[1)]. There is renewed interest in the problem at this time due to recent higher energy experiments at the CERN S$p\bar{p}$S Collider and the Tevatron at Fermilab. These experiments include measurements of

(1) $\sigma_{tot}(\bar{p}p)$
(2) Minijet production
(3) $d\sigma_{el}/dt$ and hence $B(0)$ the forward elastic scattering slope and
(4) $\rho(0)$ the ratio of real to imaginary forward amplitudes.

This last quantity is turning out to be interesting since the UA4 Group at CERN[2] find a large value $\rho(0) = 0.24 \pm .04$ at $\sqrt{s} = 546$ GeV.

We will argue that the observation of a high value of $\rho(0)$ can be naturally accomodated by QCD. At $\sqrt{s} \approx 1$ TeV average hadron collisions become semi-hard, i.e. calculable in perturbative QCD, due to the rapid increase with energy of the gluon content of the proton. In any model where the exploding gluon content of the proton drives the increase with energy of the total cross section, this essentially threshold type behaviour will result in a relatively large value of $\rho(0)$. The advent of this threshold has been seen by UA1 through the discovery of the minijet phenomena. Certain cosmic ray experiments also indicate the presence of jet-like structures produced in Multi-TeV interactions.

CENTRAL PRODUCTION AND GLUONS

Even before the discovery of copious production of gluon jets with moderate p_t (minijet phenomena) at the $Sp\bar{p}S$ the important physical role played by gluons in hadron-hadron central production was indicated by certain scaling behaviour[3]. Particles with quark structure similar to that of the proton are produced easily by fragmentation in pp interactions while those of different composition such as mesons and antibaryons are made centrally for the most part. We see in figure 1 the expression $M^3\sigma(M,s)/\Gamma$ plotted against s/M^2, where $\sigma(M,s)$ is the inclusive cross section for producing mass M in pp interactions and Γ is the width of M. This quantity behaves for central production as

$$\sigma(M,s) = f(M)F_{gg}(M^2/s) \tag{1}$$

where F_{gg} is the gluon-gluon structure function. We show two parallel curves of the naive gluon-gluon structure function with n=5

$$F_{gg}(\tau) = 9\int_\tau^1 \frac{dx}{x}(1-x)^5(1-\tau/x)^5 \tag{2}$$

This scaling is independent of the quark content of the produced resonance. Particles produced by fragmentation such as Δ^{++}, have cross sections that behave differently (see figure 1).

Figure 2 shows calculations of a simple model of $\sigma(m,s)$ versus m at fixed $\sqrt{s}$ for a series of s values for centrally produced particles. Also shown is experimental data (divided by $2J+1$). We see that the main factor determining the cross section is the mass of the state, independent of quark content. For high enough mass at a given s the cross sections flatten out to an inverse mass power behaviour. This is the region where low order perturbation theory gives reasonable production rates. It has been shown recently[4] however that B meson production (or b quark production) at $\sqrt{s} = 630$ GeV has as large an α_s^3 contribution as the lowest order α_s^2 gluon-gluon fusion result shown below.

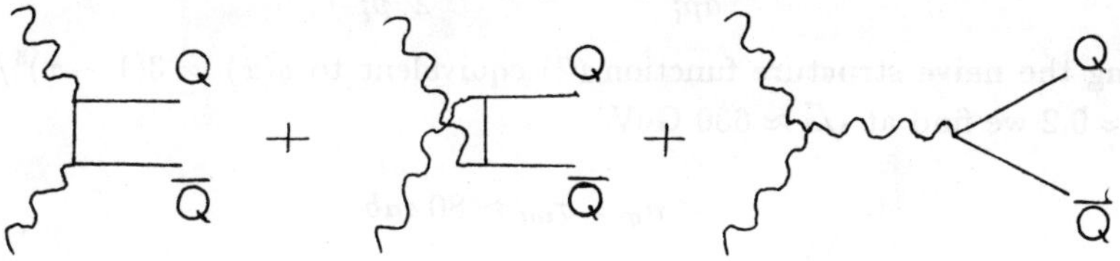

The α_s^3 contribution consists of gluon-gluon scattering followed by gluon fragmentation into a $Q\bar{Q}$ pair. Typical diagrams are

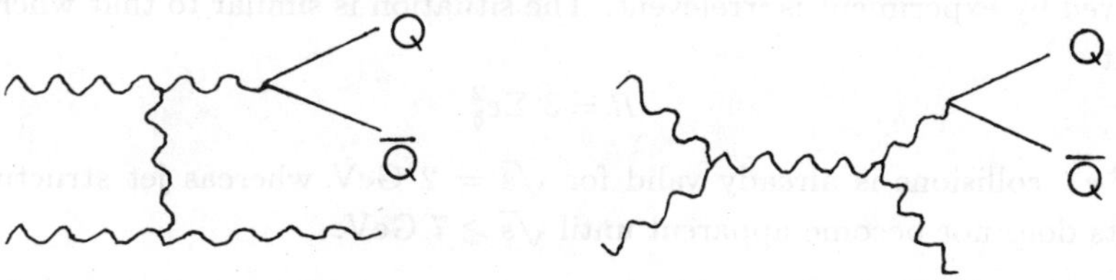

Figure 3 shows calculations of reference 4 for b production cross sections and ratios of $O(\alpha_s^3)$ and $O(\alpha_s^2)$ contributions. For charm production, the α_s^3 corrections make for a more reasonable value of quark mass $m_c = 1.5$ GeV to get agreement with measured charm cross sections (Figure 4).

In any case the centrally produced particles discussed above all are governed by the gluon-gluon structure function. We deduce below that the dominant mechanism for particle production at high enough energies is gluon-gluon scattering followed by gluon fragmentation.

GLUON JETS AND DIFFRACTION

At high enough energies (e.g. $Sp\bar{p}S$) and moderate p_t it is clear that the produced jets are gluon jets. Further we now know that the perturbative calculation of the jet cross section with transverse momentum p_t is not only valid in the hard scattering regime with $x_t = 2p_t/\sqrt{s} \approx 1$ but is in fact reliable for $p_t << \sqrt{s}$ when $p_t > \Lambda_{QCD}$ ($\Lambda_{QCD} \approx 100$ MeV). The cross section for jet production with $p_t > 1$ GeV through the dominant process $gg \to gg$ is

$$< n_g > \sigma_{tot} = \int_{1GeV} dp_t^2 \int dx_1 dx_2 g(x_1) g(x_2) \frac{d\hat{\sigma}}{dp_t^2} \tag{3}$$

with

$$\frac{d\hat{\sigma}}{dp_t^2}(gg \to gg) \approx \frac{9\pi}{2} \frac{\alpha_s^2}{p_t^4} \tag{4}$$

Using the naive structure function (2) equivalent to $g(x) = 3(1-x)^5/x$ and $\alpha_s \approx 0.2$ we find at $\sqrt{s} \approx 630$ GeV

$$< n_g > \sigma_{tot} \approx 80 \; mb \tag{5}$$

The jet cross section is of the order of the total cross section and therefore at high energy where the gluon structure is fully developed hadron collisions become perturbative or semi-hard. The fact that a 1 GeV gluon jet cannot be resolved by experiment is irrelevent. The situation is similar to that where the result

$$R = 3 \; \Sigma e_q^2 \tag{6}$$

in e^+e^- collisions is already valid for $\sqrt{s} = 2$ GeV whereas jet structure of events does not become apparent until $\sqrt{s} \geq 7$ GeV.

The jet-like behaviour of hadron-hadron interactions is first revealed by the mini-jet data of the UA1 experiment[5)] and becomes a feature of average interactions above 100 TeV apparent to even the most crude Cosmic Ray detectors[6)]. This qualitative change in the event structure is responsible for violations of Feynman and KNO scaling in the energy range $\sqrt{s} = 0.1$ to 1 TeV.

Why does this phenomenon exhibit itself as a threshold type effect? This threshold-like behaviour is associated with the rapid rise with energy of the number of relatively soft gluons once they take over the role of the valence quarks which dictate the behaviour of low energy cross sections ($\sqrt{s} \approx 100$ GeV). In the semi-hard regime the fractional gluon momenta x range from 10^{-2} to 10^{-3} typically and the gluon structure function increases faster than any power of x. A solution of the Altarelli-Parisi equations yields

$$xg(x, Q^2) \approx \exp\left(2\sqrt{\frac{3}{\pi b} \ln\left(\frac{\ln Q^2/\Lambda^2}{\ln Q_0^2/\Lambda^2}\right) \ln\frac{1}{x}} \right) \tag{7}$$

with $b = (33 - 2n_f)/12\pi$ where n_f is the number of quark flavors. The threshold behaviour is further enhanced by the rapid evolution of g(x) with Q^2 in the small x region. This explosive small x behaviour is a feature of explicit structure functions such as EHLQ[7] and DO[8]. Figure 5 shows several possible momentum distributions of gluons (the left hand side of eq. (7)) versus x for fixed Q^2 and several Q^2-independent structure functions.

We also understand where perturbation theory breaks down when p_t approaches Λ_{QCD}. Screening of the large number of soft partons stacked inside a high energy hadron will modify eq. (3) when x or p_t at a fixed energy becomes too small. This problem can be treated by introducing an eikonal formalism[1]. In such a picture the opening up of large inclusive cross sections associated with eq. (3) will make the proton blacker and result in an increase of the total cross section.

EIKONALIZATION OF GLUONS

We calculate the total hadron-hadron cross section by constructing an elastic scattering amplitude and applying the optical theorem

$$\sigma_{tot} = 4\pi Im f(s,0) \tag{8}$$

The imaginary part of the amplitude

$$Im f(s,t) = \int_0^\infty (1 - e^{i\chi(s,b)}) J_0(b\sqrt{-t}) b db \tag{9}$$

where the eikonal $\chi(s,b)$ is given by

$$\chi(s,b) = \frac{i}{2} W(b) \int_{m_0/\sqrt{s}}^1 dx_1 \int_{x_1}^1 dx_2 \int_{\delta^2}^{\hat{s}-\delta^2} d\hat{t}\, \frac{d\hat{\sigma}}{d\hat{t}}\, g(x_1,\hat{t})\, g(x_2,\hat{t}) \tag{10}$$

provided the glue-glue interaction is the only parton contribution. The parameters m_0 and δ remove the divergences of $d\hat{\sigma}/d\hat{t}$. We have here simply the QCD 2-jet cross section multiplied by a form factor $W(b)$ representing the gluon distribution in impact parameter space. We use the Fourier transform of the convoluted dipole form factors of the pp (or $\bar{p}p$) system to represent W(b)

$$W(b) = \frac{1}{8} (\mu b)^3 K_3(\mu b) \tag{11}$$

To be more realistic in terms of fitting scattering data from $\sqrt{s} \approx 20$ GeV and up we have considered more detailed models.

The following provides good fits to elastic scattering and total cross sections including the forward real-to-imaginary amplitude ratio, $d\sigma_{el}/dt$ and the forward elastic slope as a function of energy. We consider now an eikonal

$$\chi = \chi_{soft} + \chi_{gg} \tag{12}$$

where χ_{gg} is as in eq. (10) above and χ_{soft} represents soft processes and is taken to be asymptotically energy independent[1)]. At the low energy end we have a small contribution to χ_{soft} from Regge pole terms[1)]. Both contributions to χ are dependent on b through a density function of the form (11), possibly with different μ values for each term. We take the gluon distribution function

$$g(x) \approx \frac{1}{x^{1+\epsilon}}(1-x)^5 \tag{13}$$

The fits to the pp and $p\bar{p}$ data are shown in figures 6 through 9. The only parameters are the μ values, the constant describing $\chi_{soft} \approx const.\ W(b)$, the cutoffs m_0 and δ (taken equal here), Λ_{QCD} and ϵ. All of these can be deduced from other data except μ for the gluons and ϵ. We find $\epsilon \approx .1$, corresponding to the behaviour of evolved structure functions for $Q^2 \approx 10$ GeV^2 as shown in figure 5. The value of μ for the gluons comes out to be slightly smaller than μ for χ_{soft} which is taken from the electromagnetic form factor of the proton. The real part[1)] obeys

$$Ref(s,t) \approx \frac{\pi}{2} \frac{d}{dlns} Imf(s,t) \tag{14}$$

The fits shown in figure 6 yield $\rho(0) \approx .2$ at $\sqrt{s} = 546$ GeV. There is no problem in accounting for larger ρ values than predicted by previous calculations. The total cross section for $\epsilon > 0$ goes as $\sigma_{tot} \approx \log^2 s$ as $s \to \infty$ whereas for $\epsilon = 0$, $\sigma_{tot} \approx \log^2(\log s)$. The detailed form of $\hat{\sigma}_{gg}$ being slowly varying is of secondary importance. Here we have taken $\hat{\sigma}_{gg}(\hat{s}) = const.\ \theta(\hat{s} - \delta^2)$.

To calculate with confidence at very high energies, and for all values of t, one probably has to know more than we do. Certainly as we go to higher and higher energies the structure functions at smaller and smaller x come into play and these are not quantitatively established.

REFERENCES

1) Afek, Y., Leroy, C., Margolis, B. and Valin, P., Phys. Rev. Lett. 45, 85 (1980); L'Heureux, P., Margolis, B. and Valin, P., Phys. Rev. D32, 1681 (1985); Margolis, B., Valin, P. and Block, M.M., "High Energy Scenarios from Constituent Scattering", Proceedings of the IInd International Conference on Elastic and Diffractive Scattering (Rockefeller U., Oct 15-18 1987), R.L. Cool, K. Goulianos and N.N. Khuri eds.; Margolis, B., Valin, P., Block, M.M., Halzen, F. and Fletcher, R.S., "Forward Scattering Amplitudes in Semi-hard QCD", Madison Preprint MAD/PH/425 (May 1988).
2) UA4 Collaboration, Phys. Lett. 198B, 583 (1987).
3) L'Heureux, P. and Margolis, B.,"Parton Picture of Particle Production", Proceedings of the 4th Annual MRST Meeting (McGill U., May 6-7 1982), B. Margolis and T.F. Morris eds.; Gaisser, T.K., Halzen, F. and Paschos, E., Phys. Rev. D15 2572 (1977); Halzen, F. and Matsuda, S., Phys. Rev. D17, 1344 (1978).
3) Nason, P., Dawson, S. and Ellis, R.K., Fermilab preprint Pub-87/222-T (1987); Altarelli, G., Diemoz, M., Martinelli, G. and Nason, P., CERN Preprint TH 4978/88.
5) UA1 Collaboration CERN preprint EP 88-29 (1988).
6) Yamdagni, N., in "Multiparticle Dynamics 1984", G. Gustafson and C. Peterson eds. (World Scientific, Singapore, 1984); Pancheri, G. and Srivastava, Y.N., and Kim, C.S. et al. in "Physics Simulations at High Energy", V. Barger et al., eds. (World Scientific, Singapore, 1986).
7) Eichten, E.J., Hinchliffe, I., Lane, K. and Quigg, C., Rev. Mod. Phys. 56, 579 (1984).
8) Duke, D.W. and Owens, J.F., Phys. Rev. D30, 49 (1984).

Figure Captions

Figure 1: The quantity $M^3\sigma(M,s)/\Gamma$ vs s/M^2, where M is the resonance mass, $\sigma(M,s)$ the production cross section of M in pp interaction at center-of-mass energy $\sqrt{s}$ and Γ the total width of M. For references to data, see the 1980 reference 1.

Figure 2: Inclusive single-particle cross section as a function of the mass m of the produced particle at five values of $\sqrt{s}$. For references to data, see P.L'Heureux and B.Margolis, Phys. Rev. D28, 242 (1983).

Figure 3: Table of b-production cross-sections in pp, $p\bar{p}$ and $\pi^- N$ collisions (from ref.4) and ratio of the O(α_s^3) heavy quark cross section evaluated at several subtraction scales μ to the O(α_s^2) cross section as a function of the heavy quark mass m at $\sqrt{s} = 0.63$ TeV. The lowest order prediction is evaluated at a fixed value of $\mu = m$. The parton distributions of EHLQ[7)] are used.

Figure 4: Total cross section for charm production in pp collisions (from ref.4). The solid (dashed) curves determine a theoretical uncertainty band for $m_c = 1.5$ GeV (1.2 GeV).

Figure 5: The momentum distribution of gluons at small x for various evolved and non-evolved parametrizations. Labelling the curves by their values at $x = 10^{-5}$ we have in counterclockwise ordering: the naive parton model with n=5, eqn. (13) with $\epsilon = .1$, EHLQ[7)] Set 2, DO[8)] "soft", EHLQ Set 1, the latter three being evaluated at $Q^2 = 10$ GeV^2, eqn. (13) with $\epsilon = .5$ and both EHLQs at $Q^2 = 546^2$ GeV^2, which almost coincide.

Figure 6: $\rho(0)$, the forward real-to-imaginary amplitude ratio, as a function of center-of-mass energy $\sqrt{s}$ for two sets of high energy parameters: a) $\epsilon = .116$, $m_0 = 1$ GeV and the gluon $\mu = .732$ GeV, b) $\epsilon = .158$, $m_0 = 1.08$ GeV and the gluon $\mu = .809$ GeV.

Figure 7: The total cross section σ_{tot} as a function of center-of-mass energy $\sqrt{s}$ for the same sets of parameters a) and b) as in Figure 6.

Figure 8: The forward elastic slope $B(0)$ for the same sets of parameters a) and b) as in Figure 6.

Figure 9: The elastic differential cross section $d\sigma/dt$ at $\sqrt{s} = 546$ GeV for parameter set a) of Figure 6.

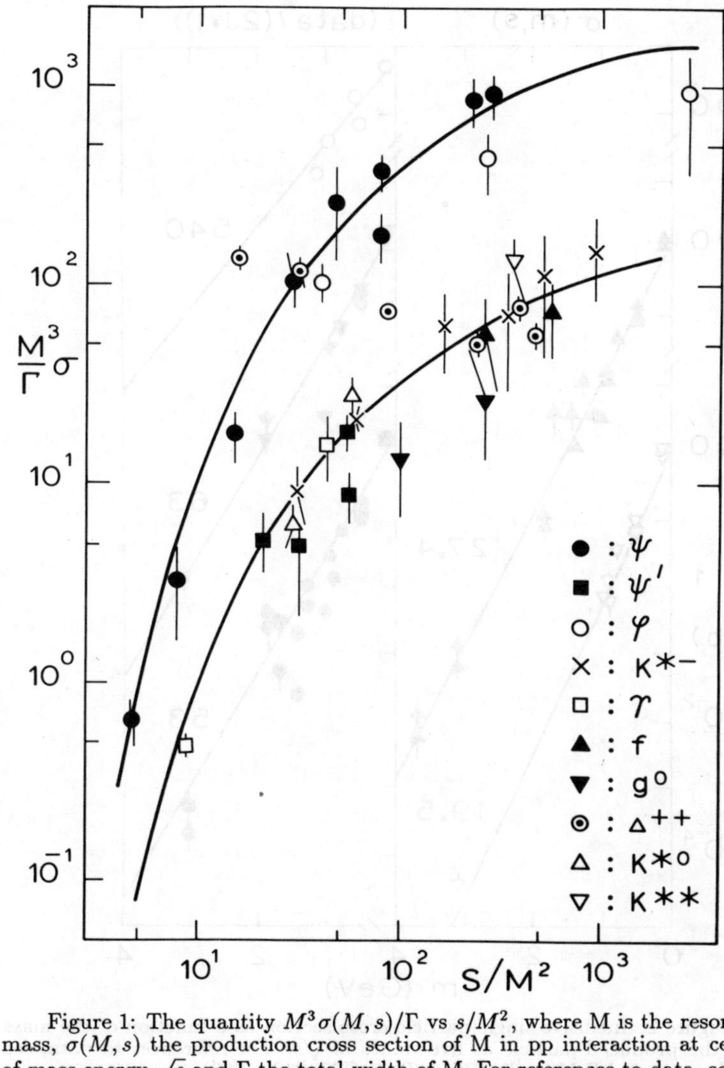

Figure 1: The quantity $M^3\sigma(M,s)/\Gamma$ vs s/M^2, where M is the resonance mass, $\sigma(M,s)$ the production cross section of M in pp interaction at center-of-mass energy $\sqrt{s}$ and Γ the total width of M. For references to data, see the 1980 reference 1.

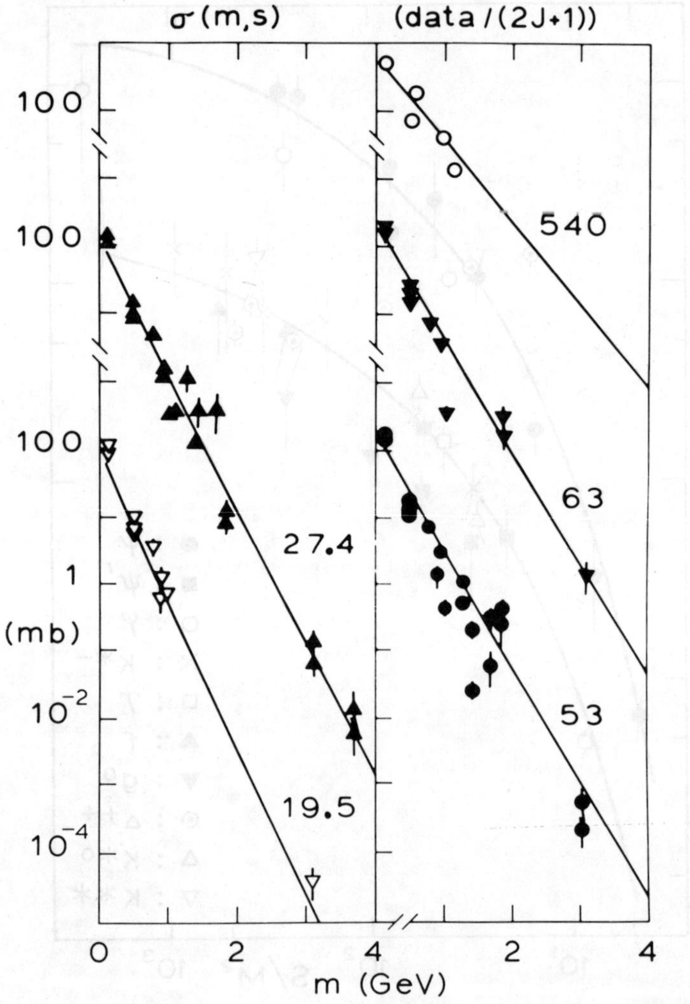

Figure 2: Inclusive single-particle cross section as a function of the mass m of the produced particle at five values of $\sqrt{s}$. For references to data, see P.L'Heureux and B.Margolis, Phys. Rev. D28, 242 (1983).

m_b(GeV)	σ	$\mu = m/2$ $\Lambda = 170$ MeV	$\mu = 2m$ $\Lambda = 170$ MeV	$\Lambda = 90$ MeV $\mu = m$	$\Lambda = 250$ MeV $\mu = m$
		$\sqrt{S} = 41$ GeV		pp	
4.5	$23 \pm^{21}_{15}$nb	40	12	13	34
5	$9.0 \pm^{8.4}_{5.9}$	16	4.7	4.9	14
		$\sqrt{S} = 62$ GeV		pp	
4.5	$142 \pm^{98}_{80}$nb	231	81	91	182
5	$66 \pm^{47}_{38}$	109	37	41	86
		$\sqrt{S} = 630$ GeV		$p\bar{p}$	
4.5	$19 \pm^{10}_{8}\mu$b	27	15	12.5	25
5	$12 \pm^{7}_{4}$	18	10	9	16
		$\sqrt{S} = 24.5$ GeV		π^-N	
4.5	$7.6 \pm^{4.7}_{3.8}$nb	12	4.6	5.2	10
5	3.1±1.5nb	4.4	1.9	2.2	3.9

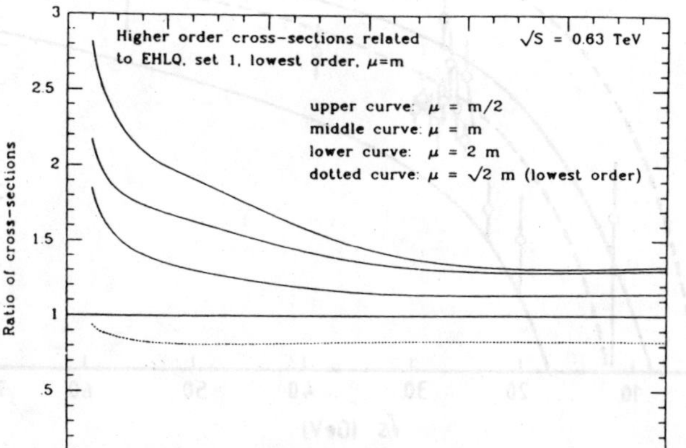

Figure 3: Table of b-production cross-sections in pp, $p\bar{p}$ and $\pi^- N$ collisions (from ref.4) and ratio of the $O(\alpha_s^3)$ heavy quark cross section evaluated at several subtraction scales μ to the $O(\alpha_s^2)$ cross section as a function of the heavy quark mass m at $\sqrt{s} = 0.63$ TeV. The lowest order prediction is evaluated at a fixed value of $\mu = m$. The parton distributions of EHLQ[7] are used.

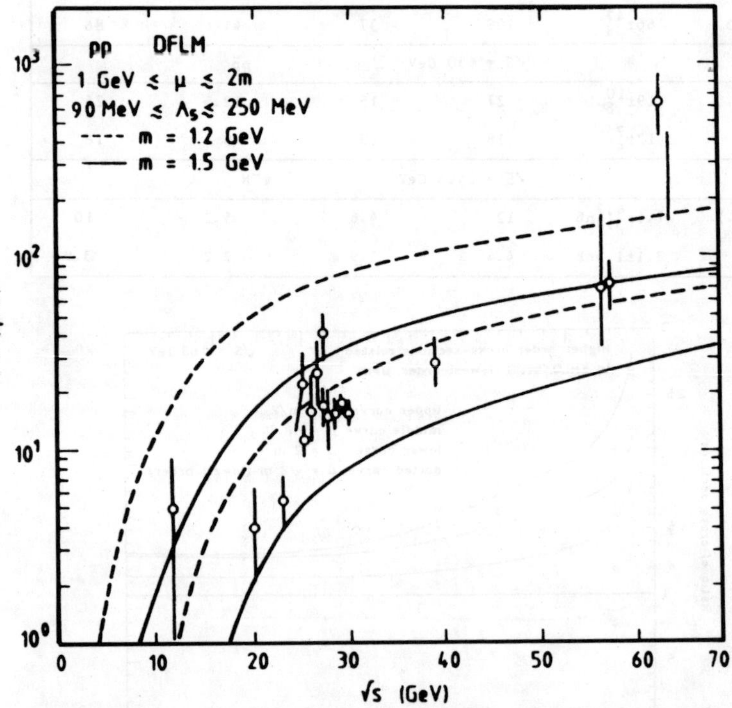

Figure 4: Total cross section for charm production in pp collisions (from ref.4). The solid (dashed) curves determine a theoretical uncertainty band for $m_c = 1.5$ GeV (1.2 GeV).

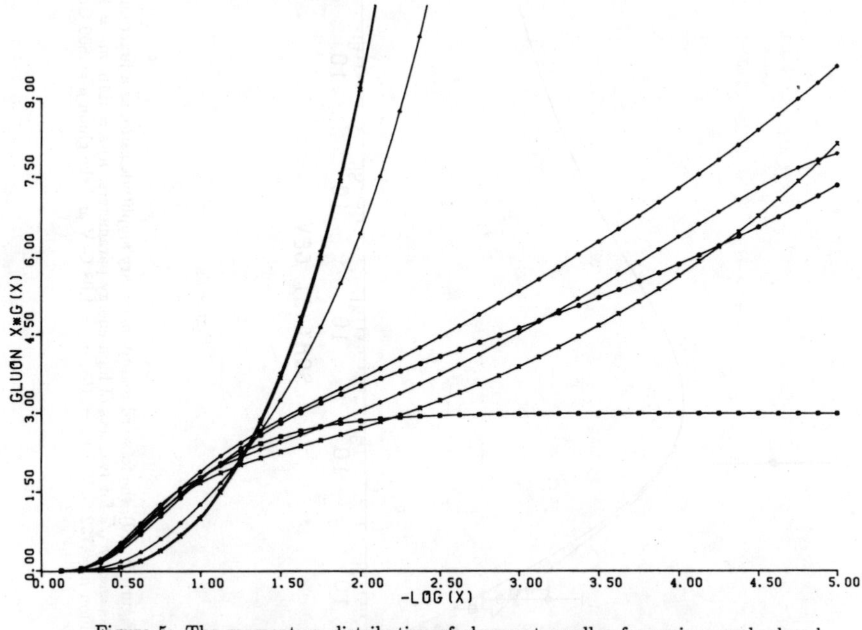

Figure 5: The momentum distribution of gluons at small x for various evolved and non-evolved parametrizations. Labelling the curves by their values at $x = 10^{-5}$ we have in counterclockwise ordering: the naive parton model with n=5, eqn. (13) with $\epsilon = .1$, EHLQ[7] Set 2, DO[8] "soft", EHLQ Set 1, the latter three being evaluated at $Q^2 = 10$ GeV^2, eqn. (13) with $\epsilon = .5$ and both EHLQs at $Q^2 = 546^2$ GeV^2, which almost coincide.

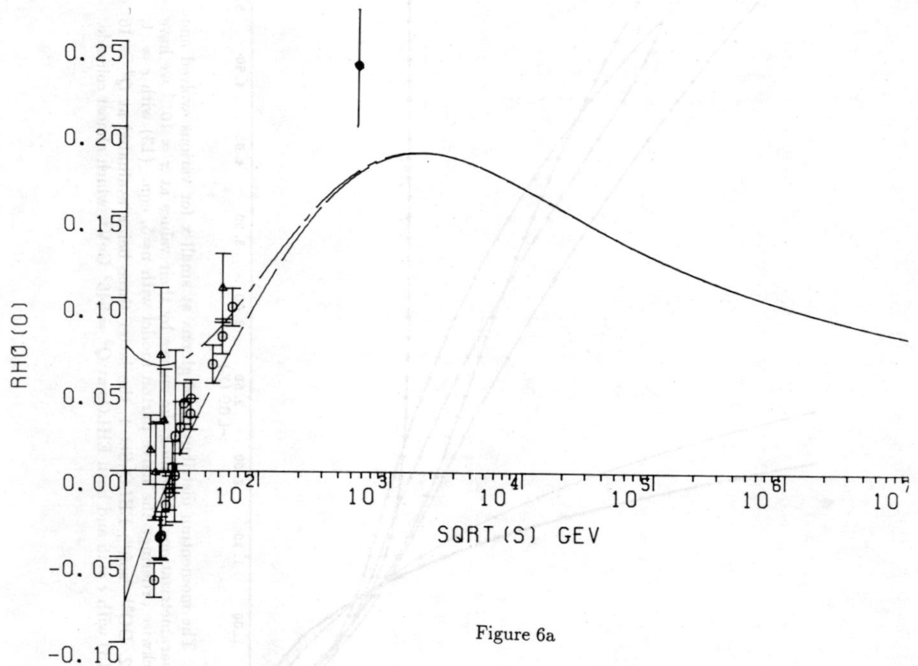

Figure 6a

Figure 6: $\rho(0)$, the forward real-to-imaginary amplitude ratio, as a function of center-of-mass energy $\sqrt{s}$ for two sets of high energy parameters: a) $\epsilon = .116$, $m_0 = 1$ GeV and the gluon $\mu = .732$ GeV, b) $\epsilon = .158$, $m_0 = 1.08$ GeV and the gluon $\mu = .809$ GeV.

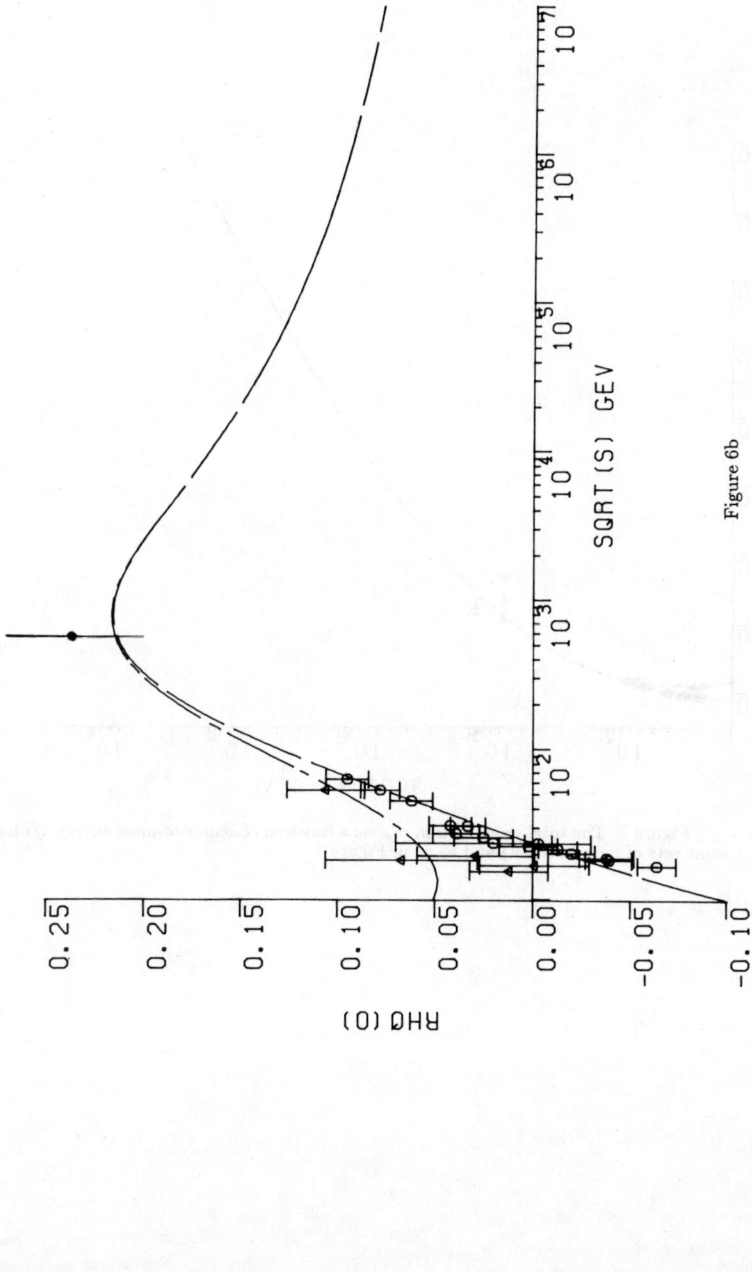
RHO (0)
0.25
0.20
0.15
0.10
0.05
0.00
-0.05
-0.10
SQRT (S) GEV
10
2
10
3
10
4
10
5
10
6
10
7

Figure 6b

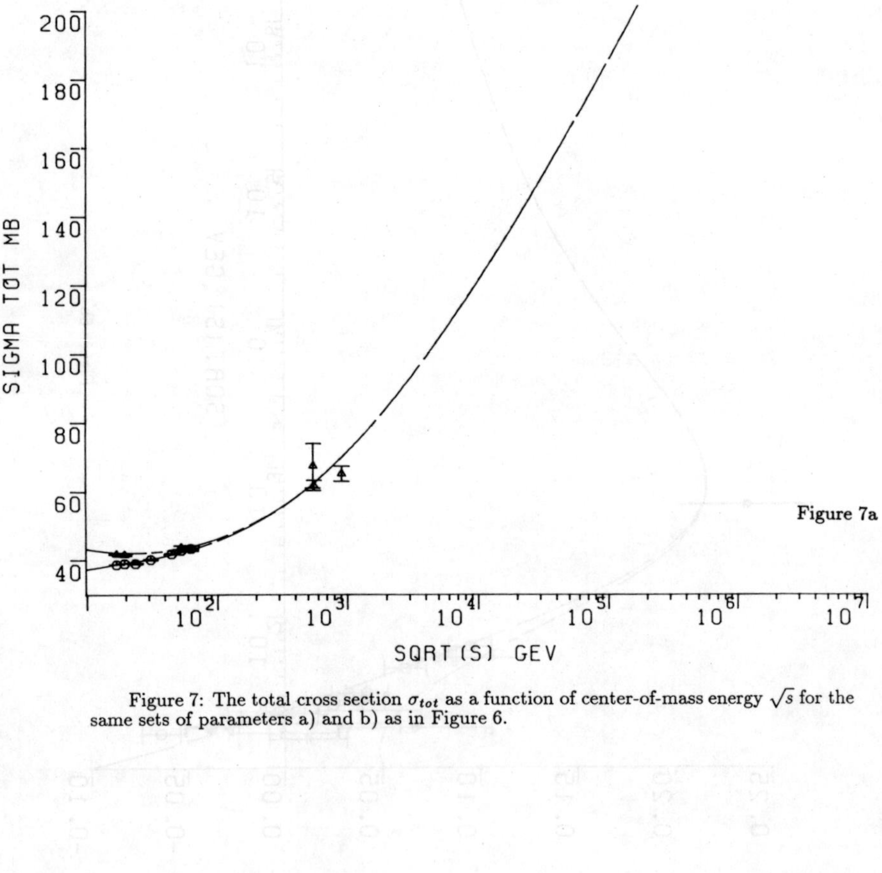

Figure 7: The total cross section σ_{tot} as a function of center-of-mass energy $\sqrt{s}$ for the same sets of parameters a) and b) as in Figure 6.

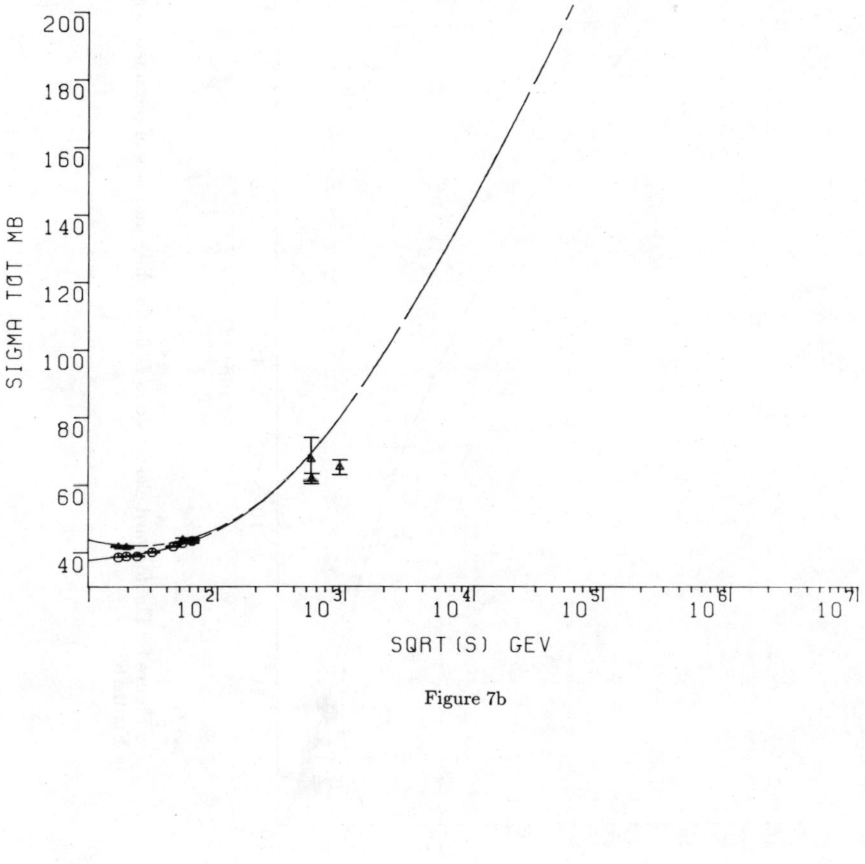
200
180
160
140
120
100
80
60
40
SIGMA TØT MB
10^2
10^3
10^4
10^5
10^6
10^7
SQRT (S) GEV

Figure 7b

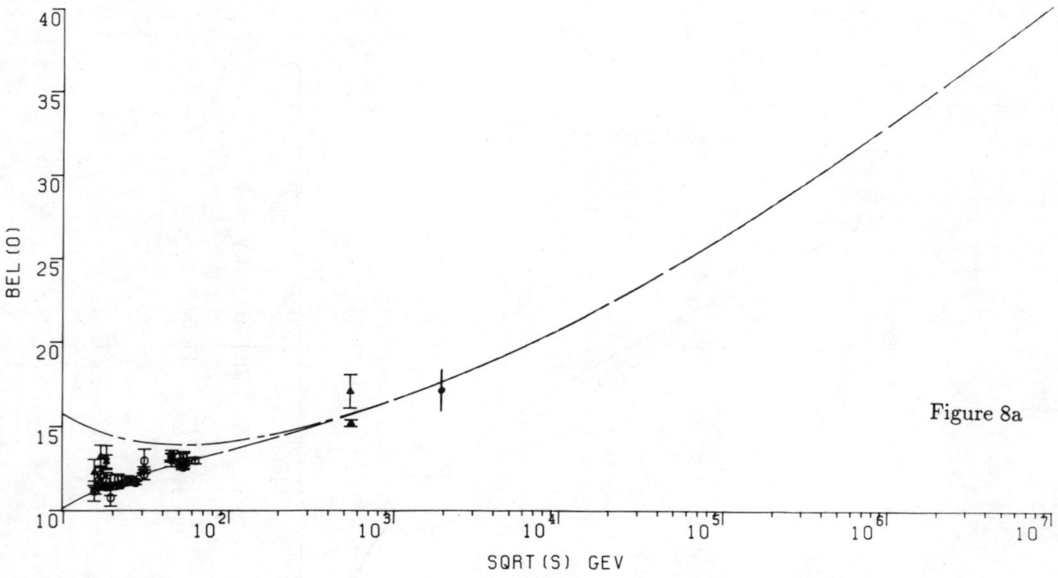

Figure 8: The forward elastic slope $B(0)$ for the same sets of parameters a) and b) as in Figure 6.

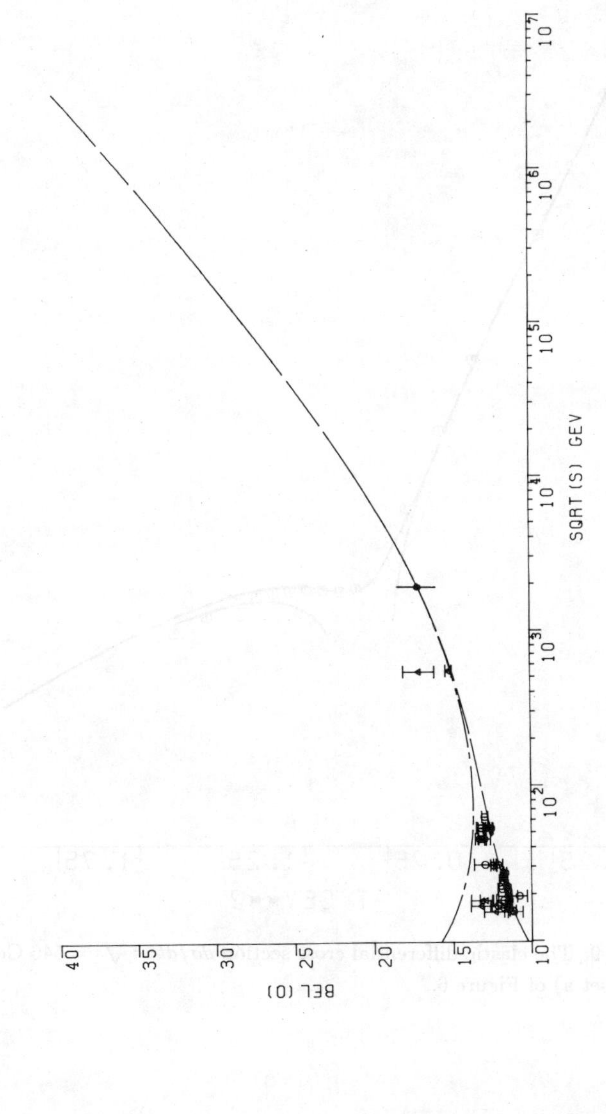

Figure 8b

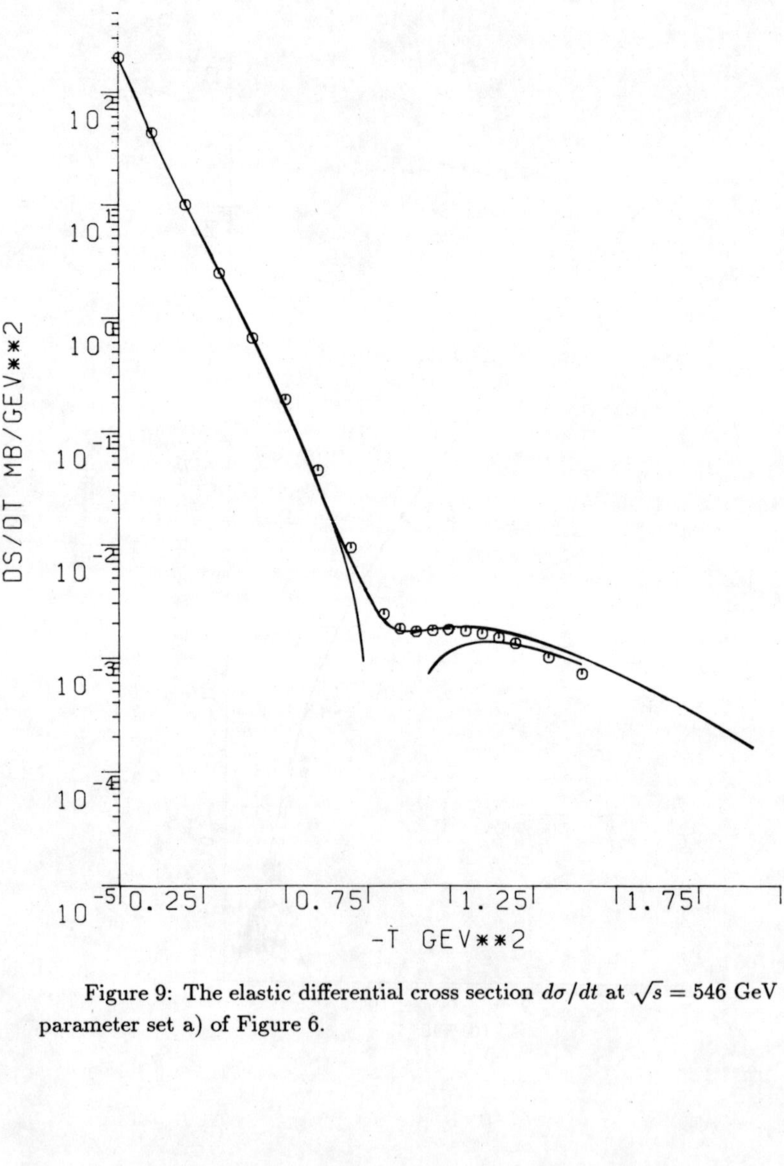

Figure 9: The elastic differential cross section $d\sigma/dt$ at $\sqrt{s} = 546$ GeV for parameter set a) of Figure 6.

CHIRAL SYMMETRY BREAKING WITH WALKING COUPLING

Bob Holdom
Department of Physics
University of Toronto
Toronto Ontario
CANADA M5S1A7

ABSTRACT

We discuss the motivation for the study of chiral symmetry breaking in gauge theories with a slowly running gauge coupling. The constant coupling case is first considered in the ladder approximation. We then demonstrate that the main qualitative features of these results also hold beyond the ladder approximation. Finally we proceed to treat the asymptotically free walking coupling case analytically in the ladder approximation.

A "walking" coupling refers to a "slowly running" gauge coupling as a function of momentum, or in other words a small β-function. There has been recent interest in such theories in the context of technicolor theories, so we will start with this motivation. In a complete technicolor theory there must be effective nonrenormalizable operators generated by new physics at a mass scale larger than the mass scale of the weak interactions. This mass scale will be referred to as the sideways scale, $\Lambda_S \gg G_F^{-1/2}$.

Among such operators are four fermion operators suppressed by Λ_S^{-2}. These come in three varieties. $\overline{T}T\overline{f}f$ operators, with T a technifermion and f a quark or lepton, in the presence of the technifermion condensate $\langle\overline{T}T\rangle$ gives rise to quark or lepton masses of order $\approx \langle\overline{T}T\rangle/\Lambda_S^2$. Second are $\overline{T}T\overline{T}T$ operators which may explicitly break certain global symmetries and thereby make a contribution to technipion masses of order $\approx \langle\overline{T}T\rangle/(F_\pi \Lambda_S)$. And third, $\overline{f}f\overline{f}f$ operators are a dangerous source of flavor changing neutral currents. Their size is proportional to Λ_S^{-2}, which in turn is determined by a fermion mass and $\langle\overline{T}T\rangle$. $\langle\overline{T}T\rangle$ was initially obtained by assuming that technicolor resembles QCD and that $\langle\overline{T}T\rangle$ is a scaled up version of $\langle\overline{q}q\rangle$. This yields the estimate $\langle\overline{T}T\rangle/\langle\overline{q}q\rangle = (F_\pi/f_\pi)^3$. The resulting value for Λ_S produces a flavor changing neutral current problem.

It is therefore of interest to find an enhancement of $\langle\overline{T}T\rangle$ relative to F_π^3. This would imply that a larger Λ_S produces the same fermion mass, and a suppression of flavor changing neutral currents would result. And note the added bonus of an increased technipion mass. We will study the extent to which technicolor has to differ from QCD so as to produce the desired enhancement of $\langle\bar{T}T\rangle$.[1-9]

$\langle\bar{T}T\rangle$ is related to the self energy function $\Sigma(p)$ appearing the technifermion propagator $S(p) = 1/(\not{p} + \Sigma(p))$:

$$\langle\bar{T}T\rangle \propto \int^{\Lambda_S} dp p^3 \frac{\Sigma(p)}{p^2+\Sigma(p)^2}. \tag{1}$$

The Λ_S cutoff appears since we are taking the expectation value of local operators which do not exist as such above the scale Λ_S. The well known asymptotic behavior of $\Sigma(p)$ in an asymptotically free theory is $1/p^2$ up to a power of a logarithm. The integral then produces no more than a logarithmic enhancement of $\langle\bar{T}T\rangle$. But suppose that instead $\Sigma(p) \approx 1/p$ over some range of $p > \Sigma(p)$. Then $\langle\bar{T}T\rangle$ will be enhanced by the ratio of mass scales over which this behavior for $\Sigma(p)$ occurs. This enhancement factor could be as large as $\approx \Lambda_S/\Sigma(0)$. We will find this $1/p$ behavior for $\Sigma(p)$ over a significant range of momentum in the case of a walking coupling.[2] We will also be concerned with finding the function of $\ln(p)$ which multiplies $1/p$.

But we may first ask how can a walking coupling occur? It is not difficult to arrange that the lowest order β-function is small by choosing a sufficient number of fermions. But lowest order perturbation theory is not sufficient to study chiral symmetry breaking. A small β-function may be obtained to all orders in a certain limit of a gauge theory: $C_2 \gg N$ and $C_2 n \ll N^3/d$ for SU(N) with n flavors and fermion representation with dimension d and Casimir C_2. Then chiral symmetry breaking occurs at small coupling $\alpha \approx 1/C_2$ and the coupling is slowly varying since $d\ln\alpha/d\ln p = \beta(\alpha)/\alpha \approx N\alpha \approx n/C_2 \ll 1$. Higher order gluonic contributions are higher order in $N\alpha$. And all fermionic contributions are suppressed because of

$C_2 n \ll N^3/d$. But this latter condition may require a number of flavors n smaller than one.

Another possibility[1] which may occur in more realistic theories is for CSB to occur in the vicinity of an ultraviolet or infrared fixed point, where by definition the β-function vanishes.

We now turn to the determination of $\Sigma(p)$. We first consider the standard Cornwall, Jackiw, and Tomboulis effective action to two loops in Landau gauge and in Euclidean space.

$$\Gamma = \frac{n}{8\pi^2}\int_0^\infty dp\, p^3 \left[\frac{4\Sigma^2(p)}{p^2+\Sigma^2(p)} - 2\ln\left[\frac{p^2+\Sigma^2(p)}{p^2}\right]\right]$$
$$- \frac{3ndC_2}{8\pi^3}\int_0^\infty dp \frac{p^3\Sigma(p)}{p^2+\Sigma^2(p)}\int_0^\infty dk \frac{k^3\Sigma(k)}{k^2+\Sigma^2(k)} f(k,p) \tag{2a}$$

The stationary condition, $\delta\Gamma/\delta\Sigma(p) = 0$, is equivalent to the Schwinger-Dyson (SD) equation in ladder approximation. Solving the latter for $\Sigma(p)$ effectively sums all ladder graphs. But the two loop Γ does not incorporate a running coupling. The latter is introduced in the so-called "improved" ladder approximation where

$$f(k,p) = \min\{1/p^2, 1/k^2\}\alpha(\max\{p,k\}) \tag{2b}$$

and $\alpha(p)$ is the running coupling. This approximation may actually be justified in the large N limit to whatever order the running coupling is calculated. But it is difficult to obtain a small β-function in a large N limit, and so we do not make use of the large N limit. Thus our use of the improved ladder approximation is not presently justified, and we will proceed beyond this approximation below.

But we shall first use the ladder approximation to study the constant coupling limit. We set

$$\alpha(k) = \alpha \quad \text{for } k \le \Lambda$$
$$\quad = 0 \quad \text{for } k > \Lambda. \tag{3}$$

The SD equation with this ultraviolet cutoff is

$$p\Sigma(p) = \frac{\alpha}{2\alpha_c}\int_0^\Lambda dk\Sigma(k)\min\{p/k,k/p\}\cdot\left[\frac{k^2}{k^2+\Sigma^2(k)}\right]. \tag{4}$$

$\alpha_c \equiv \pi/(3C_2)$ will turn out to be a critical coupling. Due to the last factor this integral equation is nonlinear. But the last factor tends to unity for $k \gg \Sigma(k)$ and acts to damp in the integral for $k \ll \Sigma(k)$. Thus we are led to linearize the equation by setting this last factor to unity and by introducing an infrared cutoff at $\kappa \equiv \Sigma(\kappa)$. We will find that $\Sigma(k)$ is not too sensitive to the exact form of the infrared damping.

We now have the two equations

$$p\Sigma(p) = \alpha/(2\alpha_c)\int_\kappa^\Lambda dk\Sigma(k)\min\{p/k,k/p\}$$
$$\kappa = \Sigma(\kappa) \tag{5}$$

from which we must determine κ as well as $\Sigma(p)$. Note that the scale of $\Sigma(p)$ will be determined even though the first equation is linear.

It is convenient to change variables,

$$x \equiv \ln(p^2/\kappa^2), \quad y \equiv \ln(k^2/\kappa^2), \quad L \equiv \ln(\Lambda^2/\kappa^2),$$
$$\varphi(x) \equiv p\Sigma(p)/\kappa^2, \tag{6}$$

which yields

$$\varphi(x) = \frac{\alpha}{4\alpha_c}\int_0^L dy\varphi(y)\exp\{-{}^1\!/_2|x-y|\},$$
$$\varphi(0) = 1. \tag{7}$$

This integral equation is equivalent to the following trivial differential equation along with two new boundary conditions.

$$\varphi''(x) + {}^1\!/_4[(\alpha/\alpha_c) - 1]\varphi(x) = 0, \tag{8}$$

$$\varphi'(0) = {}^1\!/_2\varphi(0), \quad \varphi'(L) = -{}^1\!/_2\varphi(L), \quad \varphi(0) = 1.$$

It is not difficult to check that there is no solution to this system of equations when $\alpha < \alpha_c$. But for a given $\alpha > \alpha_c$ there are sinusoidal solutions for an infinite number of L values. The different solutions have different numbers of nodes, and it is not difficult to check that the smallest L solution (which has zero nodes) corresponds to the global minimum of the original action. The boundary conditions determine the following form when the L for the zero node solution is large.

$$\varphi(x) = (L+4)/(2\pi)\sin[\pi(x+2)/(L+4)] \tag{9}$$

Identifying $\pi/(L+4)$ with the square root of the factor appearing in (8) gives

$$\Lambda/\kappa = \exp(\pi/\sqrt{(\alpha/\alpha_c)-1} - 2). \tag{10}$$

The hierarchy which develops between Λ and κ when α approaches α_c from above is of interest in the technicolor context; there Λ corresponds to the sideways scale and κ to the technifermion mass scale.

Our solution for $\Sigma(p)$ now reads

$$\Sigma(p) = (\kappa^2/p)C\sin\{[1+\ln(p/\kappa)]/C\} \quad (\kappa < p < \Lambda) \tag{11}$$

where $C = 1/\sqrt{(\alpha/\alpha_c)-1}$. Notice that $p\Sigma(p)$ peaks well above $p = \kappa$ when C is large. The integrands in the effective action behave similarly; most of the contribution to the effective action arises from momenta much larger than κ. We now understand why the solution is not too sensitive to the form of the infrared damping or to our linearization procedure. We also note that the individual terms in the action grow like C^3.

We will now try to see whether some features of this analysis holds beyond the ladder approximation. The full fermion propagator may be written in the form

$$S(p) = \{Z(p)(\not p + \Sigma(p))\}^{-1}. \quad (12)$$

Notice that we have defined $\Sigma(p)$ by extracting a factor of $Z(p)$. At this stage we need not specify a gauge; we will do so when necessary below. The full effective action becomes

$$\Gamma(S) = -\mathrm{Tr}\{\ln S^{-1}\} + \mathrm{Tr}\{(S^{-1}-\not\partial)S\} - (\text{2PI diagrams}), \quad (13)$$

where the 2PI refers to the fermion lines in these diagrams.

There is again damping in the infrared and it again proves convenient to introduce the infrared cutoff $\kappa = \Sigma(\kappa)$. This permits an expansion of $\Gamma(S)$ in powers of $\Sigma(p)/p$. If $\Sigma(p)$ behaves basically like $1/p$ as before then $\Gamma(S)$ will again be dominated by a large momentum contribution. Since this contribution appears in the Σ^2 terms, this will justify the neglect of terms higher order in Σ. Thus retaining the Σ^2 terms only is a self consistent approximation if we are able to deduce the $1/p$ behavior of $\Sigma(p)$ within this approximation.

$\Gamma(\Sigma^2)$ also involves $Z(p)$, but after we make use of the condition $\delta\Gamma(S)/\delta Z(p) = 0$ we find that most of this $Z(p)$ dependence cancels out. We obtain

$$\Gamma(\Sigma^2) = (nd/4\pi^2)\{ -c\kappa^4 + \int_\kappa dpp\Sigma(p)^2 -$$

$$- (2\alpha_c)^{-1}\iint_\kappa dpdkpk\Sigma(k)\Sigma(p)f(p,k) \}, \quad (14)$$

$$f(p,k) = Z(p)^{-1}Z(k)^{-1}(6\pi ndC_2)^{-1}\int_0^\pi d\theta \sin^2\theta\, \delta_{\alpha\beta}\delta_{\gamma\delta}K_{\alpha\beta\gamma\delta}(p,k),$$

$$K_{\alpha\beta\gamma\delta}(p,k) = \frac{\delta(\text{2PI diagrams})}{\delta S_{\alpha\beta}(p)\delta S_{\gamma\delta}(k)}\,.$$

$K_{\alpha\beta\gamma\delta}(p,k)$ is a sum of diagrams with four amputated fermion legs.

$\alpha_c = \pi/(3C_2)$ and θ is the angle between the four-vectors p and k in $K_{\alpha\beta\gamma\delta}(p,k)$. We have dropped terms independent of $\Sigma(p)$; but we have added the $c\kappa^4$ term to represent the infrared contribution to the original effective action (c is some number of order unity).

We may now use renormalization group arguments to study the scaling behavior of the Z and K functions. At this point we find that Landau gauge presents itself as the preferred gauge for the problem, as it did in the ladder approximation. The gauge parameter vanishes in Landau gauge and this turns out to be a "gauge parameter fixed point" under renormalization. That is, a term in the RG equations which accounts for the gauge parameter "evolution" appears in other gauges but vanishes in Landau gauge.

Since we are dealing with a constant coupling the β-function term in the RG equations also vanishes. Then anomalous scaling in fermion n-point functions can only occur due to a nonzero anomalous dimension $\gamma(\alpha)$ associated with renormalization of the fermion wave function. We thus have the following scaling behavior of two-point and four-point functions as we scale $p \Rightarrow \sigma p$: (as long as $(p,\sigma p) < \Lambda$ and $(p,\sigma p) \gg \kappa$)

$$Z(\sigma p) = \sigma^{-2\gamma(\alpha)} Z(p),$$

$$K_{\alpha\beta\gamma\delta}(\sigma p,\sigma k) = \sigma^{-(2+4\gamma(\alpha))} K_{\alpha\beta\gamma\delta}(p,k) \tag{15}$$

This yields

$$f(\sigma p,\sigma k) = \sigma^{-2} f(p,k). \tag{16}$$

This is naive scaling since f(p,k) has mass dimension −2. This indicates that there are no anomalous powers of (p or k)/κ. This scaling argument ignores finite mass effects and thus (16) only holds to zeroth order in $\Sigma(p)/p$. But this is all we need for f(p,k) to obtain the terms of second order in $\Sigma(p)/p$ in $\Gamma(\Sigma^2)$.

We have found that even though a zero β-function does not prevent anomalous scaling in the theory, no anomalous scaling occurs in $f(p,k)$. Note that this property depends on the way a factor of $Z(p)$ was factored out in the definition of $\Sigma(p)$: $S(p) = \{Z(p)(\not p + \Sigma(p))\}^{-1}$. This property has an important consequence for the SD equation which follows from the $\Gamma(\Sigma^2)$ in (14). Defining x and $\varphi(x)$ as before we find

$$\varphi(y) = (4\alpha_C)^{-1}\int_0^L dx\varphi(x)V(|x-y|,\alpha), \tag{17}$$

with $V(|x-y|,\alpha) \equiv pkf(p,k)$. That the kernel V is a function of $|x-y|$ follows from the fact that $pkf(p,k)$ is dimensionless and scales naively. It is therefore a function of p/k and not $(p \text{ or } k)/\kappa$.

Without knowing anything more about V it is now straightforward to show that any solution $\varphi(x)$ is either symmetric or antisymmetric about L/2. Since $\varphi(x)$ is defined as before we have therefore found that

$$\Sigma(p) \propto (\kappa^2/p)\varphi(\ln(p/\kappa)) \tag{18}$$

where the function $\varphi(\ln(p/\kappa))$ is such that $\Sigma(\Lambda) = \pm\kappa\Sigma(\kappa)/\Lambda$.

We now see that the tendency for $\Sigma(p)$ to fall like $1/p$ is much more than an artifact of the ladder approximation. We also note that the magnitude of $\Sigma(\Lambda)$ essentially confirms the enhancement of the condensate $\langle\bar{T}T\rangle$, since the condensate involves a divergent integral of $\Sigma(p)$ cutoff at Λ. Expressed in terms of $\varphi(x)$,

$$\langle\bar{T}T\rangle \propto \int_0^L dx e^{x/2}\varphi(x), \text{ where } \varphi(L) = \pm 1. \tag{19}$$

Clearly $\varphi(x)$ would have to be a very pathological function in order for this integral not to give an enhancement factor of order $e^{L/2} = \Lambda/\kappa$.

Having gained more confidence about the ladder approximation results we will return to it and consider a walking coupling. (The following analysis is also described in Ref. (6) and (8).) The lowest order running coupling normalized to be close to unity for chiral symmetry breaking is

$$\frac{\alpha(p)}{\alpha_C} = \frac{A}{\ln(p/\lambda)} \tag{20}$$

where $A \equiv 1/(b\alpha_C)$ and $p\partial_p\alpha(p) = -b\alpha(p)^2$. Notice that large A corresponds to small λ and thus a walking coupling. λ need not be associated with the confinement scale since the β-function may well change below the chiral symmetry breaking scale.

The SD equation which follows from (2a) and (2b) is

$$\psi(x) = {}^1\!/_4\int_{x_0} dy\psi(y)\exp\{-{}^1\!/_2|x-y|\}\alpha(\max\{x,y\})/\alpha_C. \tag{21}$$

$\psi(x) \equiv p\Sigma(p)/\kappa^2$ as before but now $x \equiv \ln(p^2/\lambda^2)$. The infrared cutoff is $x_0 \equiv \ln(\kappa^2/\lambda^2)$ and the $\kappa = \Sigma(\kappa)$ condition reads $\psi(x_0) = 1$. In deriving the corresponding differential equation it is convenient to express it in terms of $G(x) = \sqrt{x/x_0}\psi(x)$.[7] An additional boundary condition is also obtained.

$$G'' + \frac{1}{x(1+x)}G' + \left(-\frac{1}{4} + \frac{A-1}{2x} + \frac{2A-1}{4x^2}\right)G = 0, \tag{22a}$$

$$G(x_0) = 1, \qquad G'(x_0) = {}^1\!/_2(1+x_0)^{-1}. \tag{22b}$$

A simplification now occurs for a walking coupling. We note that x_0 is of order A since $2A/x_0 = \alpha(x_0)/\alpha_C \approx 1$. Since x_0 is the smallest x we consider, large A means that we may neglect the terms proportional to x^{-2} in (22a). The result,

$$G''(x) + \left(\frac{-1}{4} + \frac{A-1}{2x}\right)G(x) = 0, \tag{23}$$

is similar to the Whittaker equation,

$$W''(x) + \left(\frac{-1}{4} + \frac{\gamma}{x} + \frac{(1/4-\mu^2)}{x^2}\right)W(x) = 0. \tag{24}$$

But in the walking coupling approximation the two are the same since we may again neglect the x^{-2} term in the latter.

The independent solutions for G(x) are the Whittaker functions $W_{(A-1)/2,\mu}(x)$ and $W_{-(A-1)/2,\mu}(-x)$. These two solutions imply the following asymptotic behavior for $\Sigma(p)$ respectively, $p^{-2}(\ln p)^{A/2-1}$ and $(\ln p)^{-A/2}$ as $p \Rightarrow \infty$. But these two forms are well known as the asymptotic solutions to the SD equation in an asymptotically free gauge theory. The first is the regular solution and is the solution of interest since it corresponds to a dynamical mass; the second is the irregular solution and corresponds to an explicit mass. That the first is the solution to our integral equation may be seen by imposing an ultraviolet cutoff. This will give rise to an additional boundary condition upon converting to a differential equation. This in turn will require some linear combination of the above two solutions, but as the cutoff is taken to infinity the first solution is chosen.

$\Sigma(p)$ is now determined up to the infrared boundary conditions. The first condition in (22b) is $\kappa = \Sigma(\kappa)$ and serves to normalize the solution. The second is equivalent to $\Sigma'(p)\big|_\kappa = 0$ and serves to determine κ (or more precisely, κ/λ). The largest zero of $\Sigma'(p)$ is chosen since this gives a $\Sigma(p)$ with no nodes for $p > \kappa$, which in turn gives the minimum action. It is convenient to express the result in terms of the confluent hypergeometric function.

$$\Sigma(p) = \frac{\kappa^3\ U(1-A/2,1,\ln(p^2/\lambda^2))}{p^2\ U(1-A/2,1,\ln(\kappa^2/\lambda^2))} \tag{25}$$

Our solution displays a similar behavior to that found in the constant coupling case. There we found that $p\Sigma(p) = C\sin\{[1+\ln(p/\kappa)]/C$ peaked at some p » κ when the coupling was close to critical. Here we also find that pΣ(p) peaks at some p » κ when A is large. It is only for much larger p that pΣ(p) falls and approaches the standard asymptotic behavior. These results are shown in Figure (1) for two different values of A. They are compared to the numerical solutions of the nonlinear ladder SD equation without infrared cutoff. If anything, our method of linearizing and introducing the infrared cutoff yields a Σ(p) which is smaller at large p than found numerically.

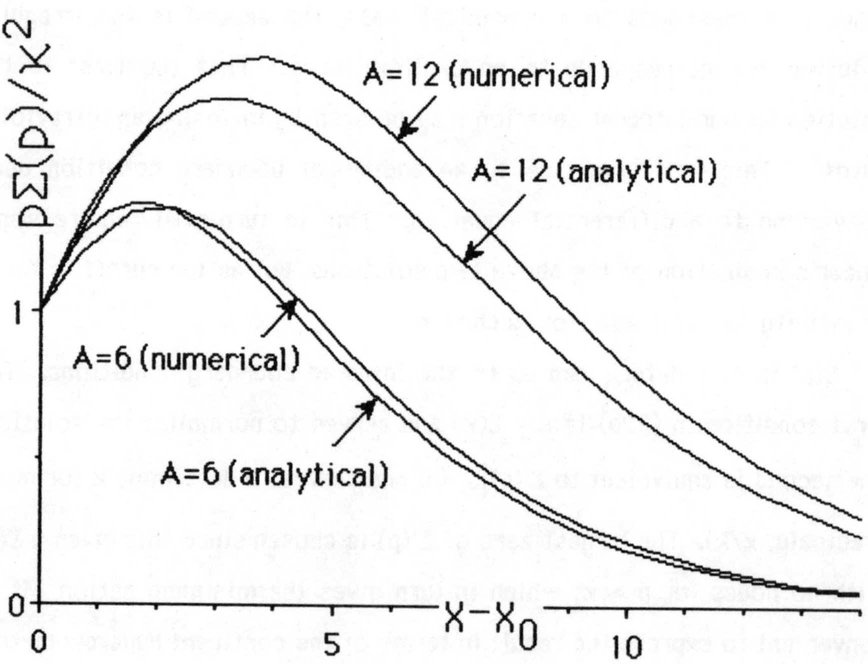

Figure (1)

Note that Σ(p) will not be significantly affected by the existence of an ultraviolet cutoff at Λ_S as long as $\Lambda_S\Sigma(\Lambda_S) \ll \kappa\Sigma(\kappa)$. On the other hand if

A is too large then this not true and the coupling is effectively constant between κ and Λ_S. The constant coupling analysis would then be more appropriate.

We return to the original motivation of this work; the enhancement that these solutions imply for the technifermion condensate $\langle\bar{T}T\rangle$. The integral for $\langle\bar{T}T\rangle$ in (1) is linearly diverging up to the scale at which $p\Sigma(p)$ starts to fall below $\kappa\Sigma(\kappa)$. The resulting enhancement is considerable for A in the range of 5 to 10. And we believe that the main prerequisite for this result, namely that $p\Sigma(p)$ is of order $\kappa\Sigma(\kappa)$ for momenta p much larger than κ, will carry over beyond the ladder approximation as was found in the constant coupling analysis.

References

1) B. Holdom, Phys. Rev. D24 (1981) 1441.

2) B. Holdom, Phys. Lett. 150B (1985) 301.

3) T. Akiba and T. Yanagika, Phys. Lett. 169B (1986) 432.

4) K. Yamawaki, M. Bando and K. Matumoto, Phys. Rev. Lett. 56 (1986) 1335; Phys. Lett. 178B (1986) 308.

5) T. Appelquist, D. Karabali and L.C.R. Wijewardhana, Phys. Rev. Lett. 57 (1986) 957;
T. Appelquist and L. C. R. Wijewardhana, Phys. Rev. D35 (1987) 774;
T. Appelquist and L. C. R. Wijewardhana, Phys. Rev. D36 (1987) 568;
T. Appelquist, D. Carrier, L.C.R. Wijewardhana, W. Zheng, Phys. Rev. Lett. 60 (1988) 1114.

6) B. Holdom, Phys. Lett. 198B (1987) 535.

7) M. Bando, T. Morozumi, H. So, K. Yamawaki, Phys. Rev. Lett. 59 (1987) 389.

8) B. Holdom and J. Terning, Phys. Lett. 200B (1988) 338.

9) B. Holdom, Phys. Rev. Lett. 60 (1988) 1233.

To B or not to B ?

Patrick J. O'Donnell
Dept. of Physics and Scarborough College
University of Toronto
Toronto, Ontario
Canada M5S 1A7

Abstract

The rare decay modes of the B meson might soon be able to test the standard model of weak interactions; these rare decays might be used to seek out a value for the top quark mass. For this to be possible, an estimate of the expected background from the standard model is needed. It is shown how reliable these estimates can be and what are the expectations of such a model.

Research Supported in part by NSERC.

Introduction.

The B(5270) meson, with $J^{PC} = 0^{-+}$ has quark content $B^+ = u\bar{b}$, $B^0 = d\bar{b}$ and two distinct anti-particles in analogy with the K meson (where $K^+ = u\bar{s}$, and $K^0 = d\bar{s}$).

An important limit on the decay of the B is that the decay into non-charm hadrons is highly suppressed, with[1]

$$\frac{\Gamma(B \to e^{\pm}\nu + \text{non-charm hadrons})}{\Gamma(B \to e^{\pm}\nu + \text{hadrons})} < 0.04.$$

This means that to a good approximation we have[2]

$$\begin{aligned}\Gamma_{TOTAL} &\approx \Gamma(b \to c + W^- \to c + X) \\ &\approx \frac{G_F^2 M_B^5}{192\,\pi^3}\, 2|V_{bc}|^2 \\ &\approx \frac{G_F^2 M_B^5}{192\,\pi^3} \left[2.95|V_{bc}|^2 + 6.3|V_{ub}|^2\right], \text{ including } b \to u.\end{aligned}$$

where V_{bc}, V_{ub} are the b,c and u,b entries in the CKM[1,3] mixing matrix.

The present limits on the CKM matrix[4] give $|V_{bc}| \approx 0.030$ to 0.062, and $|V_{ub}| \approx 0.003$ to 0.010; *i.e.*, the b quark almost, but not quite, decouples from the first family. This implies that the rare decays are important.

The rare decays of mesons and, in particular, of the K meson have been useful both in the development of the standard model and as a probe in the search for new physics. In the standard model of weak and electromagnetic interactions with three generations of quark families, - which is what I shall limit myself to in this talk - the tree-level couplings of the neutral gauge bosons are flavor diagonal. The flavor-changing, neutral-current couplings are induced at the one-loop level. These induced couplings will vanish also if the virtual quarks appearing in the loop are degenerate in mass (the GIM rule[5]). A lack of suppression will thus give a way to measure quark mass splittings. In the rare K decays, limits for the charm quark mass were inferred in just this way[6].

There are a number of reasons that point to the B meson and its rare decays to be an even better system to study than that of the K system. The lifetime of the B is of the order of 10^{-12} sec, a relatively long lifetime for such a heavy meson (remember that $m_B \approx 10m_K$). Also, in probing for a limit on the mass of the top quark, the induced neutral current coupling that links the top quark is governed by $|V_{ts}V^*_{td}|^2/|V_{us}|^2$ in K decays, which is of the order of 10^{-6}, whereas, in B decays the appropriate combination of KM matrix elements is $|V_{ts}V^*_{tb}|^2/|V_{cb}|^2$, which is of order 1. This gives us the opportunity to probe the standard model in a very different energy region. In the K system it has often proved difficult to separate the weak from the hadronic processes; in $K\bar{K}$ mixing the number of decay channels is limited by the available energy. Of course, if the phase space in the B system is much larger, there will be many more channels; the above discussion indicates that the decays are expected to be rarer in the B system but the branching ratios will be larger.

Another important consideration is that a number of laboratories are beginning to (or will in the near future) produce substantial amounts of B mesons. The existence of B factories is not too far in the future, promising copious production of B mesons - the rare decays may not be so rare after all.

The CKM Matrix - Present limits in the Standard Model

The standard model has quarks in doublets with respect to the weak interactions. These are shown below, with primes denoting the weak eigenstates of the "bottom" quarks.

$$\begin{pmatrix} u \\ d' \end{pmatrix}_L , \begin{pmatrix} c \\ s' \end{pmatrix}_L , \begin{pmatrix} t \\ b' \end{pmatrix}_L$$

These weak eigenstates are expressable in terms of the mass eigenstates by the CKM matrix, V :-

$$\Psi' = V\Psi$$

where

$$\begin{pmatrix} d' \\ s' \\ b' \end{pmatrix} = \begin{pmatrix} V_{ud} & V_{us} & V_{ub} \\ V_{cd} & V_{cs} & V_{cb} \\ V_{td} & V_{ts} & V_{tb} \end{pmatrix} \begin{pmatrix} d \\ s \\ b \end{pmatrix}$$

These can be written in terms of three angles and a phase:-

$$V = \begin{bmatrix} c_{12}c_{13} & s_{12}c_{13} & s_{13}e^{-i\delta_{13}} \\ -(s_{12}c_{13} + c_{13}s_{23}s_{13}e^{i\delta_{13}}) & (c_{12}c_{23} - s_{12}s_{23}s_{13}e^{i\delta_{13}}) & s_{23}c_{13} \\ (s_{12}s_{23} - c_{12}c_{23}s_{13}e^{i\delta_{13}}) & -(c_{12}s_{23} + s_{12}c_{23}s_{13}e^{i\delta_{13}}) & c_{23}c_{13} \end{bmatrix}$$

where $c_{ij} = \cos\theta_{ij}$ and $s_{ij} = \sin\theta_{ij}$. Empirically, $c_{ij} \approx 1$ and therefore the upper off-diagonal terms are proportional to the s_{ij}.

The typical values of the matrix elements are (in magnitude):-

$$V = \begin{bmatrix} 0.98 & 0.2 & 0.003 \\ 0.2 & 0.97 & 0.03 \\ 0.002 & 0.03 & 0.998 \end{bmatrix}$$

B-$\overline{B}$ Mixing:-ARGUS Result.

In the last year the ARGUS collaboration have presented a first measurement of the B-$\overline{B}$ mixing(7).

In the K system it has often proved difficult to separate the weak from the hadronic processes; in K$\overline{K}$ mixing the number of decay channels is limited by the available energy. In the standard model, with three generations, the B$\overline{B}$ mixing has a distinctive signature and can test the standard model or give a signal for new physics. The mixing ratio is sensitive to the mass of the t quark and imposes a constraint on m_t, the product of the bag constant "B" and f_B, $\sqrt{B}f_B$ and V_{td}. For example, $|\sqrt{B}f_B| < 200$ MeV and $m_t < 180$ GeV, $\Rightarrow |V_{td}| \geq 0.007$. Due to the lack of knowledge of δ, the CP-violating phase, a range of values of m_t can still account for the data for a reasonable choice of f_B.

What is δ?

Actually, some knowledge about δ away from the extremes of 0° or π would be of great help. The difficulty in even deciding this can be seen by considering the unitarity of the CKM matrix. For example, unitarity implies that the first and third columns should give zero for $\sum_{j=1}^{3} V^*_{3j} V_{1j}$. The angle between V^*_{ub} and $s_{12} V^*_{cb}$ is in fact δ_{13} (8).

However, it will take very accurate measurements of B-$\overline{B}$ mixing, the lifetime of the B and the angle governing the $b \to u$ transitions to settle the probable value of this CP-violating angle.

The mixing of B and $\overline{B}$ to produce dileptons of the same sign would only be sensitive to a light t-quark mass if the meson has a s quark; if the meson has a d quark there is an important dependence on the CP violating phase. In the ARGUS experiment, the decay $\Upsilon(4S) \to B^o \overline{B}^o$ has been reconstructed giving rise to a mixing parameter $r = 0.21 \pm 0.08$, where r is the ratio of same sign dileptons to opposite sign dileptons.

<u>Probing the Standard Model</u> (or background for new physics)

Why the B-system? Some reasons for the efficacy of the heavier meson system has already been given. An important reason lies in the fact that the B system will explore new, unknown parameters. In the case of the K system, with two generations there were 5 parameters, m_u, m_d, m_s, m_c and θ_{12}, all of which are now known. The addition of a new generation of t and b quarks adds 5 new parameters, of which two are known, m_b and θ_{23}, and three are still not known (m_t, θ_{13} and δ_{13}).

We can also make a comparison of the K,B systems with the other possible meson type of decays, viz., the decays of the D and θ (Toponium, whenever it is discovered[9]). These latter mesons decay with a dominant CKM matrix element $|V_{cs}|$ and $|V_{tb}|$, respectively. These are of order unity, so that the decays are fast, at least in so far as the CKM angles are concerned. In contrast, the K and B meson decays are already Cabibbo suppressed, implying that the rare processes have a relatively large branching ratio. A similar property holds in the flavor changing $|\Delta f| = 2$ mass mixing. Here the combination of CKM unitarity and the GIM mechanism $\Rightarrow \Delta m \approx 0$ if the internal quark masses are degenerate, which is the case for the D system, $\Delta m_D \approx m_s^2 - m_d^2 \approx 0$. However for the K system $\Delta m_K \approx m_c^2 - m_u^2 \neq 0$, and similarly for the B system, $\Delta m_B \approx m_t^2 - m_c^2 \neq 0$. Thus we can get information about the mass of the top quark from the Δm_B; in a similar way m_c was inferred from Δm_K[10].

CP violation.

In the standard model, CP violation occurs if all three generations are involved - in contrast with the superweak scenario. The third generation nearly decouples from the first two in the loop diagram giving Δm_K. For CP violation in the K system, $Im(\Delta m_K)$ is the appropriate part of the amplitude and is given by t exchange - thus the smallness of CP violation in the K system can be thought of as arising from the smallness of $|V_{ts}|$and $|V_{td}|$. For the B system, all three generations u,c,t can be of the same order so that CP violation can be differ from that of the K system by orders of magnitude. (If the recent reports of the measurement of the electric dipole moment of the neutron are correct, then the standard model, even with extension to more than three generations seems to be in trouble[11]).

Rare flavor-changing decays

The example that I wish to concentrate on here is the rare flavor-changing decay $b \to s\gamma$. This decay will illustrate much of the problems associated with estimating the expected standard model background. Such a process is important for the inclusive decay $B \to \gamma X_s$ and the exclusive decays such as $B \to \gamma K^*$. The process takes place at the one loop level and depends on the CKM matrix elements and the top quark mass, but does not involve the bag constant or the decay constant f_B at the quark level. At the one-loop level the branching ratio is a rapidly growing function of the top quark mass with a value of about 10^{-4} for $m_t \approx 100$ GeV. That is, we would need at least 10^6 or more decays to see a signal of about 100 events. The one loop matrix element illustrates the sort of suppression mechanism described above and how it is overcome whenever one of the quarks in the intermediate state has a mass that differs considerably from the others. The matrix element for the decay in the lowest order is

$$M = \frac{G_F}{2\sqrt{2}} \frac{e}{2\pi^2} \sum_i V_{ib} V_{is}^* F_2^i q^\mu \epsilon^\nu \bar{s} \sigma_{\mu\nu} (m_b R + m_s L) b.$$

where L(R) is the left- (right-) projection operator and the sum is over the charge +2/3 quark states u, c and t. The photon energy is q^μ and F_2^i is a function[2)] of $x_i = (m_i/M_W)^2$, where m_i is the mass of the u,c or t quark.

The function $F_2^i(x_i) \approx x_i$ for $x_i \ll 1$. Unitarity of the CKM matrix implies that the sum over the quark states in (1) may be written as

$$\sum_i V_{ib} V_{is}^* F_2^i = V_{tb} V_{ts}^* (F_2(x_t) - F_2(x_u)) + V_{cb} V_{cs}^* (F_2(x_c) - F_2(x_u)).$$

In this form the GIM mechanism can readily be seen, for in the case of all $x_i \ll 1$ the matrix elements will tend to zero. In the b decay the c- and u- quarks satisfy the condition, $x_i \ll 1$, leaving the terms involving the t quark. If the t quark has a mass of the order of the W boson mass, as seems likely, then this can form an important source of information about the value of this mass.

The expected signals for the decays of a B meson, where B=($b\bar{q}$), with q denoting a light u or d quark, are $B \longrightarrow K^* \gamma$, for a real photon[(12)] (in the rest frame of the B meson, the final state will involve a strange quark "jet" of low energy ($\approx M_B/2$) recoiling against a photon) and $B \longrightarrow K \ell^+ \ell^-$, or $K^* \ell^+ \ell^-$ although these are probably suppressed by at least a factor of $O(\alpha)$. Although the di-muon detection may be easier, with the distribution of muon pairs giving a signal for longitudinal or transverse virtual photons, the branching ratio is expected to be about 10^{-6} and less rapidly varying than in the real photon case.

Signals for new physics

There have been many papers in the last few years on possible signals for new physics; from four or more generations, SUSY particles, superstring phenomenology and so on. While these approaches tend to have a number of arbitrary parameters, it is possible that the branching ratios are significantly larger than that from the standard model. This means that we should at least have a good understanding of how well the standard model is known for a background to detect this new physics(13).

There are two major sets of problems:
(1) in exclusive decays, how well understood are the hadronization techniques, is it reasonable to use the nonrelativistic wavefunctions of the quark model and how accurately are the recoil effects known(14)?
(2) Before many of the new physics claims are made we need a good determination of the θ_{ij}, δ_{13} and $\sqrt{B}f_B$ parameters. This means that a number of good experiments on a variety of processes are required; $B\to D\ell\nu$, $B\to D^*\ell\nu$, for example to test V_{bc}, $B\to\rho\ell\nu$ to test V_{bu}, the mixing processes and the rare decays such as we have considered here, to measure the contribution of the top quark, among many.

QCD corrections

The heavy B system allows for a new regime in which to probe our understanding of perturbation theory. The weak radiative decay $b\to s\gamma$ provides a clean process to look at these 2-loop effects. Generally one expects logarithmic corrections. For $m_t \le m_W$, these need not be small, and in fact they may even dominate if $m_t \ll m_W$. These corrections have been a subject of some recent papers; although the calculations do not agree among themselves, the result appears to be an increase by about an order of magnitude (at $m_t \sim 50$ GeV) down to about 4 (at $m_t \sim 120$ GeV)[15]. (Since this talk was presented a more complete calculation[16] has shown that the increase is somewhat smaller, down by a factor of about two from these previous estimates).

Within QCD, with large recoil momentum transfers, the spectator quark will get a jolt from a hard gluon. This means that there could be a further set of diagrams with hard gluons from the one-loop process, connecting to the spectator system. These diagrams have not been calculated as yet; *a priori*, they could be as important as the diagram usually associated with flavor changing processes[17].

Conclusions

- B mesons promise a new look at the standard model in a very different energy regime. Although the decays are rare, the amplitudes can be large.

- The $B\bar{B}$ mixing results from ARGUS point to the importance of measuring f_B. An independent measurement would allow a conclusion about m_t to be made.

- Accurate measurements of the (small) V_{ub}, V_{td} and $s_{12}V_{cb}$ would give information about the CP-violating phase.

- CP violation could come about in the B system in a very different way than in the K system. This would be a first new signal of CP violation in about 25 years.

- The effects of QCD corrections enter in a different way than in the K system. The separation of short distance and long distance effects is clearer than in the light quark system.

- The effect of the QCD corrections, in the leading logarithmic approximation has been calculated in a complete way. The increase in the inclusive rate is expected to be largest for smaller top quark mass, say about 50 GeV, where the increase is about a factor of 7, down from previous estimates.

References

[1] Particle Data Group, Phys. Lett. 170B (1986) 1.

[2] B.A. Campbell and P.J. O'Donnell, Phys. Rev. D25 (1982) 1989.

[3] N. Cabibbo, Phys. Rev. Letters, 10, (1963) 531.
M. Kobayashi and T. Maskawa, Prog. Theor. Phys. 49 (1973) 652.

[4] F. Gilman, talk presented at La Thuile, Italy, March 1988.

[5] S.L. Glashow, J. Iliopoulos and L. Maiani, Phys. Rev. D2 (1970) 1285.

[6] See, for example, M.K. Gaillard, B.W. Lee and R.E. Schrock, Phys. Rev. D13, (1976) 2674.

[7] H. Albrecht *et al.*, Phys. Lett. 192B, (1987) 245.

[8] C. Hamzaoui, J.L.. Rosner and A.I. Sanda, Chicago preprint, EFI 88-07

[9] See, *e.g.*, the review by J.H. Kühn and P.M. Zerwas, Munich preprint, MPI-PAE/PTh 52/87. Also, M. Frank and P.J. O'Donnell, Toronto preprint UTPT-88-06

[10] M.K. Gaillard and B.W. Lee, Phys. Rev. D10, (1974) 897.

[11] A. Barroso, *et.al.*, Physics Letts. B196 (1987) 369; C. Hamzaoui, private communication.

[12] The present upper limit is $2.4 \cdot 10^{-4}$ for the branching ratio at 90% CL; ARGUS Collaboration (H. Albrecht *et. al.*,), DESY 88/062, May 1988.

[13] As an illustration, see the paper by R.M. Godbole, U. Türke and M. Wirbel, Physics Lett. B194 (1987) 302, for the possible enhancement in a model of extraq generations andd light Higgs particles.

[14] See, *e.g.*, T Altomari, Phys. Rev. D59 (1988) 677.

[15] S. Bertolini, F. Borzumati and A. Masiero, Phys. Rev. Lett 59, (1987) 180;

N.G. Deshpande *et. al.*, Phys. Rev. Lett 59, (1987) 183;

B. Grinstein, R. Springer and M. Wise, Phys. Lett. B202 (1988) 138.

[16] R. Grigjanis, H. Navelet, P.J. O'Donnell and M. Sutherland, Univ. of Toronto preprint UTPT-88-11.

[17] I thank J. Donoghue for discussions on this topic.

The a_1 in τ Decay

Colin Morningstar

Department of Physics
University of Toronto
Toronto, Canada M5S 1A7

The decay $\tau \to \nu_\tau \pi\pi\pi$ provides a potentially powerful means of observing the axial-vector isovector state expected in the region of 1.2 GeV. Extraction of the properties of this resonance is, however, complicated by its broad width. We examine the problems of studying such a resonance, especially the model dependence of its deduced mass and width. Within a clearly-defined and well-tested model we find $m_{a_1} = 1220 \pm 15$ MeV and $\Gamma_{a_1} = 420 \pm 40$ MeV.

I. INTRODUCTION

The structure of the three pion final states in the reaction $\tau \to \nu_\tau \pi\pi\pi$ is a potentially powerful tool for studying the axial-vector isovector a_1 state[1]. Indeed, were the a_1 a reasonably narrow resonance, the extraction of its properties from this reaction would be almost trivial: the $\tau \to \nu_\tau \pi\pi\pi$ transition scans in $m_{3\pi}$ from $3m_\pi$ to m_τ with a known probe, the axial-vector current. Problems arise, however, from ambiguities which are present in the treatment of what is actually a broad resonance. One reflection of these problems is the wide range of masses and widths which have been extracted from such data[2–5] (see Table I).

The properties of the a_1 were first studied in hadronic reactions. These studies[6,7] were complicated not only by the broad width of the a_1, but also by the presence of strong backgrounds, especially the diffractive "Deck effect"[8]. The standard values of the a_1 parameters quoted by the Particle Data Group[9] are based mainly on these analyses. The discrepancy between the properties of the a_1 deduced from τ decay and the hadronic reactions is obviously a cause for concern. This discrepancy has been recently studied by Bowler[10] who, based on an observation in Ref. 3, found that: 1) the different results quoted by Refs. 2, 3 and 4 (Ref. 5 was published later) were due primarily to their differing treatments of the effects of the large a_1 width and not to differences in their data, and 2) that a fit to the data allowing for unknown mass dependence of the decay amplitudes off resonance gave an a_1 mass, but perhaps not a width, from τ decay consistent with that from the hadronic experiments. He also showed that a similar mass dependence was indicated by fits to the ρ-dominated processes $e^+e^- \to \pi^+\pi^-$ and $\tau \to \nu_\tau \pi\pi^0$. The result is a very strong phenomenological case that there is no major discrepancy between the τ and hadronic data.

Following the observation made in Ref. 3 of the crucial role mass dependence plays in extracting the properties of the a_1 from τ decay, we began this work with the aim of deducing this mass dependence from theory. To the extent that our assumptions can be trusted (a subject we address below), this eliminates a large part of the uncertainty in deducing the a_1's parameters. We also examine the importance of the $\rho\pi$ D-wave in the decay $a_1 \to \rho\pi$, as well as the effects in $\tau \to \nu_\tau \pi\pi\pi$ due to the radial excitation of the pion.

II. TIME-ORDERED vs. COVARIANT PERTURBATION THEORY

If hadrons were pointlike particles, there would be no problem (at least in principle) in treating the decay $\tau \to \nu_\tau \pi\pi\pi$ with arbitrary precision in terms of the hadronic fields. Since hadrons are instead composite objects, one encounters difficulties in using them as effective degrees of freedom. Perhaps foremost among these problems are those which emerge when one tries to perform covariant perturbation theory.

A Feynman diagram is a sum of time-ordered graphs, some of which involve particles, and others antiparticles, propagating forward in time. A crucial requirement for the graphs of time-ordered perturbation theory to combine into a covariant graph is that the elementary vertices be pointlike so that, for example, $e^- \to e^-\gamma$ and $e^+e^- \to \gamma$ have the same strength. Hadrons do not possess this property. Figure 1 illustrates this point in the context of the problem at hand. The $a_1 \to \rho\pi$ vertex of Figure 1(a) is something that can be measured on shell; it can also be computed with a modest degree of reliability in various models since it involves the creation of only a single quark-antiquark pair. The creation of $a_1\rho\pi$ by the strong interaction from the vacuum, the amplitude at the heart of the Z-graph of Figure 1(b), is not easily measurable nor is it an amplitude one expects to match in strength the $a_1 \to \rho\pi$ decay amplitude since it involves the creation from the vacuum of three $q\bar{q}$ pairs. In fact, there are cases, such as $\gamma p \to p$ versus $\gamma \to p\bar{p}$ at low energy, where one knows from direct measurements that the Z-graph vertex is suppressed.

This observation creates a dilemma: one can either have a manifestly covariant calculation containing a piece of manifestly incorrect physics or one can have a calculation which uses the correct vertices but is not manifestly covariant. For narrow resonances, this dilemma is not a serious one because each second-order Z-graph such as Figure 1(b), representing a fraction $|(E-M)/2E|$ of the total Feynman graph for pointlike coupling, is already unimportant since in the resonance peak $|(E-M)/2E| \approx \Gamma/2M \ll 1$. However, the a_1 populates the 3π mass spectrum over essentially its entire kinematic range. Thus, $|(E-M)/2E|$ can be substantial and the standard use of covariant perturbation theory is probably unjustified.

As a consequence of this, and given our more or less complete ignorance of everything about the Z-graph vertices except that they seem likely to be suppressed[11], we *ideally* would like to keep in our analysis only the naively time-ordered graphs, *i.e.*, the non-Z graphs such as Figure 1(a). They are the graphs which resonate while the Z-graphs provide a smooth non-resonant "background". We could then calculate $\tau \to \nu_\tau \pi\pi\pi$ via the non-Z graphs corresponding to sequential processes like $\tau \to \nu_\tau a_1$, $a_1 \to \rho\pi$, $\rho \to \pi\pi$ and $\tau \to \nu_\tau \pi'$, $\pi' \to \rho\pi$, $\rho \to \pi\pi$. For several reasons, one of which is the desire to represent the effects of Z-graphs, we would then add to these resonant graphs a parametrized non-resonant amplitude. *Realistically*, however, the convenience of the covariant approach is difficult to sacrifice, especially when dealing with four-body phase space. Therefore, we choose instead to restore covariance to the amplitude, but we do so taking care that 1) the resonant portion of the amplitude is exactly the same as that given by the non-Z time-ordered diagrams, and 2) the parametrization of the non-resonant background amplitude is sufficiently rich that it can cancel the spurious effects of the Z-graphs induced by making the resonant amplitude part of a covariant amplitude.

The conceptual separation of the resonant (and presumably dominant) time-ordered graphs from non-resonant effects, in addition to being important for the reasons just given, has other advantages. One is that the off-shell behaviour of the non-Z graphs is (more or less) completely calculable. This eliminates one of the most serious uncertainties in studying the a_1 in τ decay. Another is that all of the allowed $a_1 \to \rho\pi$ and $\rho \to \pi\pi$ couplings for the resonant graphs can be treated. This contrasts with the covariant graph procedure in which most of the allowed couplings are eliminated without full justification.

III. VERTICES

Consider the vertices of Figure 2. The most general $a_1 \to \rho\pi$ vertex factor is

$$\Gamma^{\mu\nu}(a_1 \to \rho\pi) \propto \alpha g^{\mu\nu} + \beta(k-p)^\mu(q+p)^\nu + \gamma(k+p)^\mu(q+p)^\nu + \delta(k-p)^\mu(q-p)^\nu + \epsilon(k+p)^\mu(q-p)^\nu, \tag{1}$$

where $q = p + k$ and α, β, $\ldots\epsilon$ are form factors which can depend on q^2 and k^2. In the usual treatment, only α and β are allowed to be non-zero[12]. In general, this is correct *only* if all of the participants in the decay $a_1 \rightarrow \rho\pi$ are on mass shell since then $\epsilon_{a_1}\cdot q = \epsilon_\rho\cdot k = 0$, where ϵ_{a_1} and ϵ_ρ are polarization vectors. Similarly, the general $\rho\rightarrow\pi\pi$ vertex factor may be written

$$\Gamma^\nu(\rho\rightarrow\pi_2\pi_3) \propto f(p_2 - p_3)^\nu + g(p_2 + p_3)^\nu, \tag{2}$$

and the usual neglect of g is valid only when the ρ is on mass shell. For the $\pi'\rightarrow\rho\pi$ vertex factor one has

$$\Gamma^\nu(\pi'\rightarrow\rho\pi) \propto a(q + p)^\nu + b(q - p)^\nu. \tag{3}$$

Finally, one must also allow for mass dependence in the weak $W\rightarrow a_1$ and $W\rightarrow\pi'$ vertices,

$$\Gamma^{\sigma\mu}(W \rightarrow a_1) = -if_{a_1}(q^2)g^{\sigma\mu}, \tag{4}$$

$$\Gamma^\sigma(W \rightarrow \pi') = -f_{\pi'}(q^2)q^\sigma. \tag{5}$$

We thus see that the unknown off-shell behaviour of these graphs can be problematical: not only does one not know the dependence of f_{a_1} and $f_{\pi'}$ on q^2 (the dependence associated with the usual parametrization[10] of the off-shell behaviour), but one also knows neither the off-shell dependence of α, β, f and a nor the functions γ, δ, ϵ, g and b which can play a role off shell.

Given all of these problems, one might despair of computing the correct mass dependence in $\tau \rightarrow \nu_\tau\pi\pi\pi$. However, almost all of the ambiguities we have just discussed are associated with the non-resonant background which according to the discussion of time-ordered perturbation theory above is an unknown common to all analyses anyway. There is no unique way of expressing the resonant behaviour of these vertex factors in a covariant fashion, as one might suspect from, *e.g.*, the fact that Eqn. (1) has five amplitudes while the on-shell vertex has only two. However, it is obviously convenient to choose an implementation which, in some sense, keeps unwanted Z-graphs associated with the resonant graph to a minimum. One might[13–16], for instance, take $\gamma = \delta = \epsilon = 0$ in Eqn. (1). While certainly an acceptable alternative, this choice produces a vertex $\Gamma^{\mu\nu}(a_1\rightarrow\rho\pi)$ which can annihilate pseudoscalar components of an off-shell a_1 and create scalar components in an off-shell ρ. This means that the non-resonant parts of the covariant amplitudes would appear even in channels with the "wrong" quantum numbers with respect to our resonant time-ordered amplitude. We therefore choose the route of constraining the vertex factors $\Gamma^{\mu\nu}(a_1 \rightarrow \rho\pi)$ and $\Gamma^\nu(\rho \rightarrow \pi_2\pi_3)$ so that they may produce only transversely polarized vector particles *even off shell.* This choice has the additional advantage that its vertices annihilate the $k^\mu k^\nu$ part of the vector propagators which are relatively complicated for broad, off-shell particles. Applying these constraints, the strong vertex factors become

$$\Gamma^{\mu\nu}(a_1 \to \rho\pi) = f_{a\rho\pi}(q^2,k^2)\left(-g^{\mu\nu} + \frac{k^\mu k^\nu}{k^2} + \frac{q^\mu q^\nu}{q^2} - \frac{k\cdot q}{k^2 q^2} q^\mu k^\nu\right)$$
$$+ g_{a\rho\pi}(q^2,k^2)\left(\frac{k\cdot q}{q^2}q^\mu - k^\mu\right)\left(q^\nu - \frac{k\cdot q}{k^2}k^\nu\right), \qquad (6)$$

$$\Gamma^\nu(\rho \to \pi_2\pi_3) = -if_{\rho\pi\pi}(k^2)(p_2^\nu - p_3^\nu), \qquad (7)$$

$$\Gamma^\nu(\pi' \to \rho\pi) = if_{\pi'\rho\pi}(q^2,k^2)\left(q^\nu - \frac{k\cdot q}{k^2}k^\nu\right). \qquad (8)$$

The form factors can now be calculated using the quark model.

We exploit the quark model by using the relativized mock meson method[17,18]. The basic idea of the method is that, in the weak binding limit, the quark model provides a Lorentz invariant decomposition of any matrix element which can be put in one-to-one correspondence with a physical matrix element of interest. One can therefore associate the true Lorentz invariant form factors with quark model formulas valid in the weak binding limit. The mock meson method then either uses these formulas or some relativized version of them with physical parameters. One thus assumes that the formulas, or some relativized version of them, can be extrapolated from the weak binding limit ($\mathbf{p}^2/m^2 \ll 1$) to the physical case where in fact $\mathbf{p}^2/m^2$ is of order unity.

In general, the q^2-dependence of the weak form factors is unknown. However, as explained above, we wish to use the q^2-dependence appropriate to the time-ordered resonant graphs. In the approximation where only the direct $W \to a_1$ and $W \to \pi'$ processes contribute, this dependence is trivial since the time-ordered graphs just involve the off-energy-shell but otherwise physical intermediate states and hence the on-mass-shell matrix elements

$$\langle a_1(\mathbf{p}_a, s_a)|A^{\dagger\mu}(0)|0\rangle = f_{a_1}\epsilon^{*\mu}(\mathbf{p}_a, s_a), \qquad (9)$$

$$\langle \pi'(\mathbf{p}_{\pi'})|A^{\dagger\mu}(0)|0\rangle = -if_{\pi'}p^\mu_{\pi'}, \qquad (10)$$

where f_{a_1} and $f_{\pi'}$ are *constants* and $A^\mu(0)$ is the axial-vector current. There are, it should be noted, higher order graphs which can give energy dependence to these form factors. If, for example, we relax the assumption that the a_1 is excited directly by the W but still maintain the assumption of resonance dominance (which should be reliable), then the a_1 could be excited indirectly via a process like $W \to a_1' \to a_1$. Here, the a_1' is a radial excitation of the a_1 and the $a_1' \leftrightarrow a_1$ mixing can occur via common strong (virtual or real) decay channels. Since such a mixing is unlikely to be strong and given that it would in the first approximation be energy independent, it should be safe to ignore corrections to the f_{a_1} = constant approximation.

The weak decay constants can be estimated in the mock meson method which gives

$$f_{a_1} = \frac{4\sqrt{2\pi}\tilde{M}_{a_1}^{1/2}}{(2\pi)^{3/2}} \int_0^\infty dp\, p^2 R_{a_1}(p) \left(\frac{p}{E}\right)\left(\frac{m}{E}\right)^\alpha, \tag{11}$$

$$f_{\pi'} = \frac{-4\sqrt{3\pi}}{(2\pi)^{3/2}\tilde{M}_{\pi'}^{1/2}} \int_0^\infty dp\, p^2 R_{\pi'}(p) \left(\frac{m}{E}\right)^{1+\beta}, \tag{12}$$

where $\phi_{nlm}(\mathbf{p}) \equiv (-i)^l R_{nl}(p) Y_{lm}(\Omega_\mathbf{p})$ is the unit normalized momentum-space wavefunction, m is the constituent quark mass and $E = (\mathbf{p}^2 + m^2)^{1/2}$. We determine the wavefunctions by diagonalizing the nonrelativistic Coulomb plus linear potential problem in a harmonic oscillator basis with the harmonic oscillator parameter used as a variational parameter. The mock meson masses $\tilde{M}$ are both taken to be 1.3 GeV. The factors of $(m/E)^x$ are inserted to allow for relativistic corrections, as suggested in Ref. 17. Table II shows the results obtained from (11), (12) and the analogous formula for f_π as a function of the parameter x. Since relativistic corrections are expected to be more important in the pion than in the a_1, it is not surprising that the pion favours a larger value for x. For the fits we perform, we take $f_{\pi'}/f_{a_1} = 0.2\ \text{GeV}^{-1}$.

The strong form factors $f_{a\rho\pi}$, $g_{a\rho\pi}$ and $f_{\rho\pi\pi}$ can be computed by similar methods. Since they involve the creation of an additional $q\bar{q}$ pair, they are less trivial dynamically than the weak form factors which just involve the action of a current. As previously mentioned, we compute them in the flux tube breaking model[19] in the $^3P_0$ limit[20]. Since, once again, we wish to build in the mass dependence appropriate to the time-ordered resonant graphs, we study (6)-(8) for physical states, where we have

$$\langle \rho(p_\rho s_\rho)\pi(p_\pi)|H_{sb}(0)|a_1(p_a s_a)\rangle = -i\epsilon_\nu^*(p_\rho s_\rho)(f_{a\rho\pi} g^{\nu\mu} + g_{a\rho\pi} p_a^\nu p_\rho^\mu)\epsilon_\mu(p_a s_a), \tag{13}$$

$$\langle \rho(p_\rho s_\rho)\pi(p_\pi)|H_{sb}(0)|\pi'(p_{\pi'})\rangle = -f_{\pi'\rho\pi}\epsilon_\mu^*(p_\rho s_\rho)p_{\pi'}^\mu, \tag{14}$$

$$\langle \pi(p_2)\pi(p_3)|H_{sb}(0)|\rho(p_\rho s_\rho)\rangle = f_{\rho\pi\pi}\epsilon_\mu(p_\rho s_\rho)(p_2 - p_3)^\mu. \tag{15}$$

The form factors are then obtained by equating the flux tube breaking amplitudes to the expressions (13)-(15) and replacing mass-squares by the appropriate Dalitz plot variables. Some numerical results are given in Table III.

IV. PROPAGATORS

With the "bare" vector-meson propagator $G_0^{\mu\nu}$ given by

$$-iG_0^{\mu\nu}(k) = \frac{-g^{\mu\nu} + k^\mu k^\nu/m_0^2}{k^2 - m_0^2 + i\epsilon}, \tag{16}$$

where m_0 is the "bare" mass, and the one-particle-irreducible bubble $\omega_{\mu\nu}$ of the general form

$$\omega_{\mu\nu}(k) = -i\left[\omega_1(k^2)(-g_{\mu\nu} + k_\mu k_\nu/k^2) - \omega_2(k^2)k_\mu k_\nu/k^2\right], \tag{17}$$

one finds, since $G^{-1} = G_0^{-1} - \omega$, that the propagator for the interacting a_1 or ρ is given by

$$-iG_{\mu\nu}(k) = \frac{-g_{\mu\nu} + \left(\dfrac{k^2 - \omega_1(k^2) + \omega_2(k^2)}{m_0^2 + \omega_2(k^2)}\right)k_\mu k_\nu/k^2}{k^2 - m_0^2 - \omega_1(k^2) + i\epsilon}. \tag{18}$$

With our use of transverse vertices, the $k^\mu k^\nu$ term conveniently never enters. For the π', the full propagator is of course analogous with the numerator in (18) equal to unity. As usual we write

$$\omega_1(k^2) = m^2(k^2) - m_0^2 - im_R\Gamma_R(k^2), \tag{19}$$

so that the denominators are in the Breit-Wigner form

$$k^2 - m^2(k^2) + im_R\Gamma_R(k^2). \tag{20}$$

We define the mass m_R of the resonance to be the value of $m(k^2)$ when $m^2(k^2) = k^2$ and the on-resonance width to be $\Gamma_R(m_R^2)$.

If only the non-Z time-ordered graphs are summed, one obtains a noncovariant "resonant" propagator whose denominator is given by

$$2E_R\left[E - E_R - \Delta(E) + i\frac{\tilde{\Gamma}_R(E)}{2}\right], \tag{21}$$

where $E_R = \sqrt{\mathbf{k}^2 + m_R^2}$, $E = \sqrt{\mathbf{k}^2 + k^2}$, $\tilde{\Gamma}_R(E)$ is the mass-dependent width related to the covariant width by $\tilde{\Gamma}_R(E) = (m_R/E_R)\Gamma_R(k^2)$ and $\Delta(E)$ is the (renormalized) level shift satisfying $\Delta(E_R) = 0$.

The mass shift functions are related to the widths via

$$m_n^2(s) = m_n^2 - \frac{1}{\pi}P\int_{s_{th}}^{\infty} ds'\left[\frac{m_n\Gamma_n(s')}{s'-s} - \frac{m_n\Gamma_n(s')}{s'-m_n^2}\right], \tag{22}$$

$$\Delta_n(E) = -\frac{1}{2\pi}P\int_{E_{th}}^{\infty} dE'\left[\frac{\tilde{\Gamma}_n(E')}{E'-E} - \frac{\tilde{\Gamma}_n(E')}{E'-E_n}\right]. \tag{23}$$

For practical reasons, we have restricted the sum over channels to $\rho\pi$ and $K^*\bar{K} + \bar{K}^*K$ in calculating the widths.

V. RESULTS AND DISCUSSION

The differential decay rate for $\tau \to \nu_\tau \pi\pi\pi$ is the sum of the two Feynman diagrams shown in Fig. 3 and given by

$$\frac{d^3\Gamma_{\tau\to\nu_\tau 3\pi}}{ds\, ds_1\, ds_2} = \frac{G_F^2 \cos^2\theta_c m_\tau^3}{16\pi^2 s}\left(1-\frac{s}{m_\tau^2}\right)^2 \left[\left(1+\frac{2s}{m_\tau^2}\right)\rho_a(s,s_1,s_2)+\rho_p(s,s_1,s_2)\right], \quad (24)$$

where $s=(p_1+p_2+p_3)^2$, $s_1=(p_2+p_3)^2$, $s_2=(p_1+p_3)^2$ and $s_3=(p_1+p_2)^2$ and ρ_a and ρ_p are axial-vector and pseudoscalar spectral densities. These densities can be written

$$\rho_a(s) = f_{a_1}^2 P_{a_1}(s) F_{a_1}(s_1,s_2), \quad (25)$$

$$\rho_p(s) = s f_{\pi'}^2 P_{\pi'}(s) F_{\pi'}(s_1,s_2), \quad (26)$$

where

$$P_n(s) = \left[\left(s-m_n^2(s)\right)^2 + m_n^2\Gamma_n^2(s)\right]^{-1}, \quad (27)$$

and

$$F_n(s_1,s_2) = 2(8\pi)^{-3}\left[P_\rho(s_1)D_n(s_1,s_2)+P_\rho(s_2)D_n(s_2,s_1)+2\Delta(s_1,s_2)I_n(s_1,s_2)\right], \quad (28)$$

in which

$$\Delta(s_1,s_2) = P_\rho(s_1)P_\rho(s_2)\left[(s_1-m_\rho^2)(s_2-m_\rho^2)+m_\rho^2\Gamma_\rho(s_1)\Gamma_\rho(s_2)\right], \quad (29)$$

and the width of the ρ-meson is given by

$$\Gamma_\rho(s_i) = \Gamma_\rho(m_\rho^2)\left(\frac{m_\rho}{\sqrt{s_i}}\right)\left(\frac{f_{\rho\pi\pi}(s_i)}{f_{\rho\pi\pi}(m_\rho^2)}\right)^2\left(\frac{s_i-4m_\pi^2}{m_\rho^2-4m_\pi^2}\right)^{3/2}. \quad (30)$$

The direct (D) and interference (I) factors are functions of the Dalitz plot variables and include the strong decay form factors. The total mass-dependent widths appearing in (27) are given by

$$m_{a_1}\Gamma_{a_1}(s) = \int ds_1 ds_2 F_{a_1}(s_1,s_2) + m_{a_1}\Gamma_{a_1 K^* K}(s), \quad (31)$$

$$m_{\pi'}\Gamma_{\pi'}(s) = \int ds_1 ds_2 F_{\pi'}(s_1,s_2). \quad (32)$$

The a_1 width is shown in Fig. 4 as a function of s; the mass shift function given by Eqn. (22) is shown in Fig. 5.

Numerical integrations over the Dalitz plot variables s_1 and s_2 can be done at this stage either partially to produce projections corresponding to s_1 distributions as a function of s or fully to produce the three pion mass spectrum. Finally, in order to compare our results with experiment, the theoretical results must be convoluted with a detector resolution function.

As the discussion of Section II emphasizes, we must allow, in addition to the a_1 and π' resonances expected in this region, a general non-resonant background amplitude. However, an initial fit to the data without such background indicates that the resonant terms are dominant. As a result, we needn't be very sophisticated in our parametrization of the background amplitude: any scheme that can represent small, smooth effects will suffice. We proceed on two fronts:

1) We modify the suspect Z-graph terms present in the covariant (but now by choice transverse) a_1 amplitude. Given the quality of the initial resonant fit, we are satisfied to do this by considering modifications to just the denominator of the covariant a_1 propagator. Our procedure is to decompose

$$\Pi \equiv (s - m(s)^2 + im_{a_1}\Gamma_{a_1}(s))^{-1} = \Pi_{res} + \Pi_{nonres} \tag{33}$$

where

$$\Pi_{res} = [2m_{a_1}(\sqrt{s} - \tilde{m}(s)) + im_{a_1}\Gamma_{a_1}(s)]^{-1} \tag{34}$$

is the purely resonant time-ordered piece of the propagator Π in the rest frame of the a_1 and $\tilde{m}(s)$ is the mass function due to coupling with only those states involved in the resonant propagator. We then study the effect on the deduced a_1 parameters from the replacement

$$\Pi \longrightarrow \Pi(\alpha) \equiv \Pi_{res} + \alpha\Pi_{nonres}. \tag{35}$$

2) We add a polynomial background term. Such a term allows us to take into account many possible small effects, including transverse Z-graph effects that cannot be described simply by $\alpha \neq 1$, non-transverse Z-graph effects, contributions from other channels (*e.g.*, the low mass tails of radial excitations of the a_1 and ρ and the high mass tail of the pion) and residual experimental backgrounds.

With some prejudices based on the discussions in Section II, we define a "preferred fit" shown in Figure 6 with the following ingredients:

1) An a_1 with properties as described above but with $\alpha = 0$ in its propagator (see Eqn. (35)) and with the mass shift given in (23) ignored. The choice $\alpha = 0$ is suggested but not required by the fits (see below); we prefer it since Z-graph suppression is expected on the grounds discussed above. (It should be noted that $\alpha = 0$ corresponds to Z-graph suppression in the a_1 rest frame and cannot strictly be interpreted as effecting such a

suppression in general.) The decision to neglect the s-dependence of the mass shift arises from our inability to calculate it with confidence; see the discussion below. We also ignore the mass shift in the ρ-meson propagator.

2) A π' with the properties predicted using the relativized quark model and the flux tube breaking model. The addition of a π' is not required by our fits, but we include it in our "preferred fit" because it is unequivocably expected with properties we are convinced cannot be too far from those we use. The total width appearing in the π' propagator contains, in addition to the partial width for $\pi' \to (\pi\pi)_\rho\pi$, a constant allowing for the decay $\pi' \to (\pi\pi)_S\pi$ via the broad $\pi\pi$ S-wave; the value of this constant is taken to be 150 MeV.

3) A polynomial background used to absorb various small residual effects described above which might otherwise distort the deduced a_1 properties.

While we prefer this fit, we use it mainly as the starting point for an exploration of uncertainties, trying not to allow our preferences to have much influence on our conclusions about the ranges in which the true a_1 parameters lie.

The discussion of the reliability of our results is not a simple matter since many of the uncertainties in our conclusions arise from theory, not experiment. An important ingredient of our analysis, as emphasized in Sections II and III, is the isolation of the time-ordered resonant piece of the usual Feynman graph and our subsequent analysis of its off-mass-shell behaviour. One consequence of this is that the weak decay constant f_{a_1} for the resonant part of the Feynman graph is approximately independent of $m^2_{3\pi}$. Bowler's analysis[10] shows that this conclusion can be reached (within errors) on phenomenological grounds by fitting the data; our arguments provide a rationale for considering the mass dependence of f_{a_1} to be known and hence reduce this source of error on the fitted a_1 mass and width.

However, a significant number of other uncertainties remain. The situation is summarized in Table IV where we show the results and qualities of various fits:

1) "preferred": This is the fit described above. It always gives acceptable confidence levels and no other fits we try ever provide any significant improvements.

2) $\alpha \to 1$: This fit explores the importance of the suppression of Z-graphs due to setting $\alpha = 0$ in the preferred fit (see Eqn. (35)). We note that χ^2 is a very flat function of α. (The $\alpha = 0$ value is preferred by the ARGUS data by about one standard deviation). The fitted a_1 parameters are, however, insensitive to this change, the effect being absorbed into the background polynomial coefficients c_n.

3) $m \to m(s)$: This fit includes the mass function $m_{a_1}(s)$ appearing in the full a_1 propagator. We consider our computed $m_{a_1}(s)$ to be considerably less reliable than *e.g.*, our $\Gamma_{a_1}(s)$. This is simply because such a mass function depends on the coupling of the resonance to all other channels, whereas our formulas include only the coupling to $\rho\pi$ and $K^*\bar{K}+\bar{K}^*K$. When only the $\rho\pi$ channel is considered, this mass function increases rather dramatically through the resonance. The coupling to $K^*\bar{K}+\bar{K}^*K$ considerably lessens this effect and we expect higher channels to further flatten this function near the a_1 resonance. These uncertainties lead us to exclude this effect from our "preferred" fit, but as Fit 3 shows, the data alone cannot exclude the possibility that such effects are important. (Note that for simplicity we have ignored SU(3) symmetry breaking effects on the relative strength of the $K^*\bar{K}$ and $\rho\pi$ couplings.) Our inability to reliably predict $m_{a_1}(s)$ represents, in our opinion, the greatest source of uncertainty in extracting the a_1 mass and width from the data[21].

4) $c_n = 0$ (no polynomial background): Here we see that some sort of non-resonant amplitude is required by the data, as the confidence level of the fit with the coefficients c_n set to zero drops dramatically. This is to be expected. Note that the a_1 parameters do not shift very much despite this large decrease in the quality of the fit.

5) $f_{\pi'} = 0$: The absence of the radial excitation of the pion with the properties predicted by our model has essentially no effect on our conclusions concerning the parameters of the a_1. However, if we were to allow a large $\pi'(1300)$ contribution, our conclusions might be affected. This possibility could be ruled out experimentally by studying the 3π Dalitz plot.

6) $f^D_{a\rho\pi} = 0$: Our model predicts the ratio of the D- and S-wave amplitudes in $a_1 \to (\pi\pi)_\rho\pi$ to be $A[a_1 \to (\rho\pi)_D\pi]/A[a_1 \to (\rho\pi)_S\pi] = -0.15$. Since the addition of the predicted D-wave produces theoretical $\pi\pi$ projections of the 3π Dalitz plot in better agreement with experiment, as indicated in Figures 7, we have little reason to worry that this model dependence is adversely affecting our conclusions. We also have the known success of the model for the analogous D/S ratio in $B \to \omega\pi$ to support our confidence. This fit shows that our conclusions on the a_1 mass and width are in any event insensitive to this ratio. The value of D/S extracted from a fit to the experimental Dalitz plot projections is -0.14 ± 0.03 which agrees well with our model's prediction.

7) $\beta \to 0.3$ GeV: The mass dependence of the hadronic form factors is already well-tested by studies of strong decay processes. For example, the $\pi\pi$ partial widths of the natural parity sequence $\rho(770)$, $f_2(1270)$, $\rho_3(1690)$, $f_4(2030)$, ... are well-described by our model even though they contain relative phase space factors $(p/p_0)^{2L+1}$ which are very sensitive to the intrinsic scale p_0 of the transition[19]. Also, the model actually predicts

with surprising accuracy the absolute width of the a_1. There are, nevertheless, reasons to be skeptical: the apparently similar electromagnetic form factors computed with the same wavefunctions are too hard; *i.e.*, they drop too slowly with q^2. We have accordingly considered the effects of varying the form factor slope parameter β from 0.4 GeV to 0.3 GeV in this fit. The χ^2 change indicates that, as in the global fit, smaller β values are not preferred.

8) No $K^*\bar{K}+\bar{K}^*K$: Not only does the $K^*\bar{K}+\bar{K}^*K$ channel have a dramatic effect on the mass shift function, but also, as this fit shows, its inclusion in the total mass-dependent a_1 width can alter the deduced resonance parameters.

On considering all of these uncertainties, we are led to the conclusions given in the last row of Table I. First, note the consistency of our mass and width[22] for the a_1 with the standard values[9] based on hadronic reactions[6,7]. Secondly, the observation that $f_{a_1} \neq 0$ reveals the failure of the "wavefunction at the origin" approximation. Since the weak current creates the $u\bar{d}$ pair in the a_1 at a point, f_{a_1} is also proportional to $\psi(0)$ and vanishes in the $L = 1$ states in the nonrelativistic limit. It has been appreciated for some time that non-leading terms in $\mathbf{p}/m$ give a substantial value to the a_1 weak decay constant[18]. The value we extract is consistent with such expectations (see Table II). It is also roughly consistent with expectations from current algebra sum rules[23].

VI. CONCLUDING REMARKS

A higher-statistics study of $\tau \rightarrow \nu_\tau \pi\pi\pi$ could improve our understanding of the a_1 in several ways. It would, first of all, directly reduce the errors in the a_1 parameters, although with a reduction in experimental errors by a factor of two these would become dominated by theoretical uncertainties. The availability of meaningful data at higher 3π mass would indirectly reduce theoretical uncertainties by providing a measurement of backgrounds (both axial-vector and pseudoscalar) off resonance which could be extrapolated under the resonance. Very high quality data might also see or limit the effect of the opening of the $K^*\bar{K} + \bar{K}^*K$ channel for a_1 decay, thereby eliminating a further theoretical uncertainty. There are also, unquestionably, improvements that can be made in our theoretical understanding of these decays including careful study of the presumed small residual mass dependence of f_{a_1}, direct calculation of the Z-graph suppressions, consideration of higher mass virtual channels, the study of theoretical and phenomenological constraints on the momentum dependence of strong form factors, etc. We believe that such experimental and theoretical efforts are well-justified and could lead to a substantial improvement in our knowledge of the a_1 and π'. However, we must also acknowledge that the large width of the a_1 presents a barrier to a precise determination of its properties that, at least for now, remains.

VII. ACKNOWLEDGEMENTS

This work was done in collaboration with N. Isgur and C. Reader from the University of Toronto. We gratefully acknowledge the hospitality of the Department of Theoretical Physics, Oxford, where much of this work was completed. We are particularly indebted to M. G. Bowler for conversations.

REFERENCES

[1] N. Isgur, C. Morningstar and C. Reader, to be published.

[2] W. B. Ruckstühl *et al.* (the DELCO collaboration), Phys. Rev. Lett. **56**, 2132 (1986).

[3] W. B. Schmidke *et al.* (the MARK II collaboration), Phys. Rev. Lett. **57**, 527 (1986).

[4] H. Albrecht *et al.* (the ARGUS collaboration), Z. Phys. **C33**, 7 (1986).

[5] H. R. Band *et al.* (the MAC collaboration), SLAC preprint SLAC-PUB-4333 (1987).

[6] C. Daum *et al.*, Nucl. Phys. **B182**, 269 (1981).

[7] J. Dankowych *et al.*, Phys. Rev. Lett. **46**, 580 (1981).

[8] R. T. Deck, Phys. Rev. Lett. **13**, 169 (1964).

[9] Particle Data Group, Phys. Lett. **B170**, 1 (1986).

[10] M. G. Bowler, Phys. Lett. **B182**, 400 (1986).

[11] There are occasions on which general principles such as gauge or chiral invariance dictate that graphs other than the non-Z graph must contribute; so far as we are aware, this is not one of them.

[12] In most of the experimental analyses listed in Table I, the stronger assumption that only α is non-zero was made. The form factors γ, δ, ϵ and g have been more or less universally ignored: see Refs. 13-16.

[13] W. R. Fraser, J. R. Fulco and F. R. Halpern, Phys. Rev. **136**, B1207 (1964).

[14] G. Dillon and M. M. Giannini, Nuovo Cimento **54A**, 937 (1968).

[15] H. Kuhn and F. Wagner, Nucl. Phys. **B236**, 16 (1984).

[16] H. E. Haber and G. L. Kane, Nucl. Phys. **B129**, 429 (1977); B. J. Read, Nucl. Phys. **B64**, 511 (1973).

[17] S. Godfrey and N. Isgur, Phys. Rev. **D32**, 189 (1985).

[18] C. Hayne and N. Isgur, Phys. Rev. **D25**, 1944 (1982); for a more recent version of this calculation, see Ref. 17.

[19] R. Kokoski and N. Isgur, Phys. Rev. **D35**, 907 (1987).

[20] The flux tube breaking model of Ref. 19 is for many purposes, including those of this paper, very similar in its consequences to the $^3P_0$ quark pair creation model. See, for the latter, A. LeYaouanc, L. Oliver, O. Pene and J. C. Raynal, Phys. Rev. **D8**, 2223 (1973); Phys. Rev. **D9**, 1415 (1974); Phys. Rev. **D11**, 1272 (1975); M. Chaichan and R. Kogerler, Ann. Phys. (N.Y.) **124**, 61 (1980).

[21] The importance of the mass function $m(s)$ has recently been considered in N. A. Törnqvist, Z. Phys. **C36**, 695 (1987). We find an effect that is considerably weaker than the one found in this reference, perhaps because of its use of pointlike form factors. We also have a serious qualitative disagreement in that we find that the effect of the mass increasing through the resonance is to decrease, rather than increase, the intrinsic width $\Gamma(m^2)$.

[22] It should be noted that the hadronic reactions were fit with a far less elaborate a_1.

[23] S. Weinberg, Phys. Rev. Lett. **18**, 507 (1967); H. J. Schnitzer and S. Weinberg, Phys. Rev. **164**, 1828 (1967).

Table I: Masses and Widths of the a_1

Source	Mass (MeV)	Width (MeV)
DELCO (τ decay)[2]	1056±20±15	$476^{+132}_{-120}\pm 54$
MARK II (τ decay)[3]	1194±14±10	462±56±30
ARGUS (τ decay)[4]	1046±11	521±27
MAC (τ decay)[5]	1166±18±11	405±75±25
$\pi^- p \to \pi^- \pi^+ \pi^- p$ [6]	1280±30	300±50
$\pi^- p \to \pi^- \pi^0 \pi^+ n$ [7]	1240±80	380±100
PDG[9]	1275±28	316±45
Bowler[10]	1235±40	400±100
this work	1220±15	420±40

Table II: The Weak Decay Constants*

	$x = 0$	$x = 1/2$	$x = 1$	experimental value
f_π (GeV)	0.19	0.14	0.11	0.132
f_{a_1} (GeV^2)	0.22	0.16	0.12	0.25 ± 0.02
$f_{\pi'}$ (GeV)	0.08	0.04	0.02	

*f_{a_1} and $f_{\pi'}$ are calculated using (11) and (12); f_π is determined using an equation analogous to (12). The parameter x corresponds to α and β in (11) and (12), respectively. The experimental value quoted for f_{a_1} is determined from a fit to the absolute rate for $\tau \to \nu_\tau \pi\pi\pi$.

Table III: The Strong Decay On-Shell Form Factors

	Model Prediction	Fit to Data
$f_{\rho\pi\pi}(m_\rho^2)$	6.08†	6.08 ± 0.04
$f_{a\rho\pi}(m_{a_1}^2, m_\rho^2)$ (GeV)	4.8	4.6 ± 0.2‡
$g_{a\rho\pi}(m_{a_1}^2, m_\rho^2)$ (GeV)	6.0	5.4 ± 0.5‡
$f_{\pi'\rho\pi}(m_{\pi'}^2, m_\rho^2)$	5.8	—
$f_{aK^*K}(m_{a_1}^2, m_{K^*}^2)$ (GeV)	7.3	—
$g_{aK^*K}(m_{a_1}^2, m_{K^*}^2)$ (GeV)	11.9	—

†Fit to $\rho \to \pi\pi$ to determine string breaking constant γ_0.

‡Fit to $\tau \to \nu_\tau \pi\pi\pi$ (Fig. 6) and $\pi\pi$ projections (Fig. 7).

Table IV: Various Fits to the ARGUS[4], MARK II[3] and DELCO[2] Three-Pion Mass Spectra

	ARGUS			MARK II			DELCO		
Fit	m_{a_1} (GeV)	Γ_{a_1} (GeV)	C.L.	m_{a_1} (GeV)	Γ_{a_1} (GeV)	C.L.	m_{a_1} (GeV)	Γ_{a_1} (GeV)	C.L.
1) "preferred"†	1.213±.011	0.434±.030	0.54	1.25±.05	0.58±.10	0.51	1.18±.06	0.43±.19	0.12
2) $\alpha \to 1$	1.219±.010	0.396±.024	0.37	1.24±.03	0.49±.07	0.56	1.19±.06	0.42±.15	0.14
3) $m \to m(s)$	1.236±.012	0.349±.019	0.36	1.26±.04	0.41±.05	0.36	1.22±.07	0.37±.12	0.16
4) $c_n \to 0$	1.242±.011	0.487±.025	0.04	1.25±.02	0.51±.06	0.24	1.22±.04	0.57±.13	0.18
5) $f_{\pi'} \to 0$	1.215±.011	0.434±.029	0.54	1.25±.04	0.57±.09	0.50	1.18±.06	0.43±.18	0.12
6) $f^D_{a\rho\pi} \to 0$	1.212±.011	0.436±.029	0.56	1.25±.04	0.58±.10	0.48	1.18±.06	0.43±.19	0.11
7) $\beta \to 0.3$ GeV	1.203±.008	0.376±.022	0.15	1.22±.02	0.47±.06	0.63	1.19±.05	0.48±.14	0.18
8) no $K^*\bar{K} + \bar{K}^*K$	1.207±.009	0.421±.026	0.53	1.21±.03	0.49±.08	0.27	1.17±.05	0.37±.17	0.10
best estimate*	1.220±.015	0.400±.045		1.24±.04	0.49±.09		1.20±.06	0.46±.16	

†A description of this fit is given in Section V. Each of the subsequent fits differs from Fit 1 only by the change indicated in the first column.

*We quote here values based on the distinct fits 1), 2), 3), 4) and 7) weighted by both the statistical confidence levels of the fits and to some extent our confidence in the physics of the fits. The numbers quoted are therefore somewhat subjective.

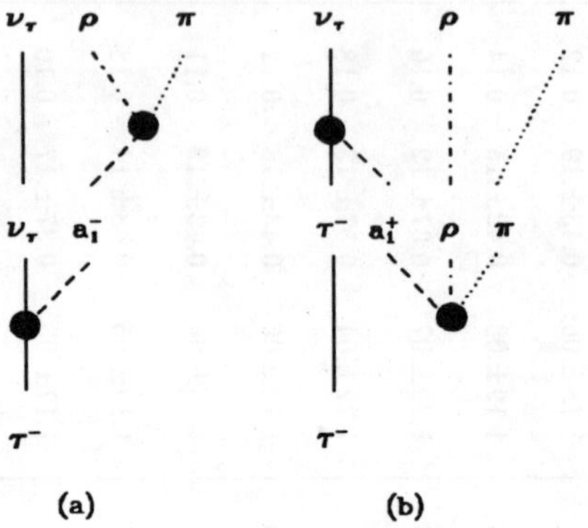

Figure 1: Two time-ordered graphs which would combine to form a covariant Feynman graph for pointlike particles: (a) strong decay following the weak creation of a_1^-, (b) an $a_1\rho\pi$ "vacuum fluctuation" followed by annihilation of a_1^+.

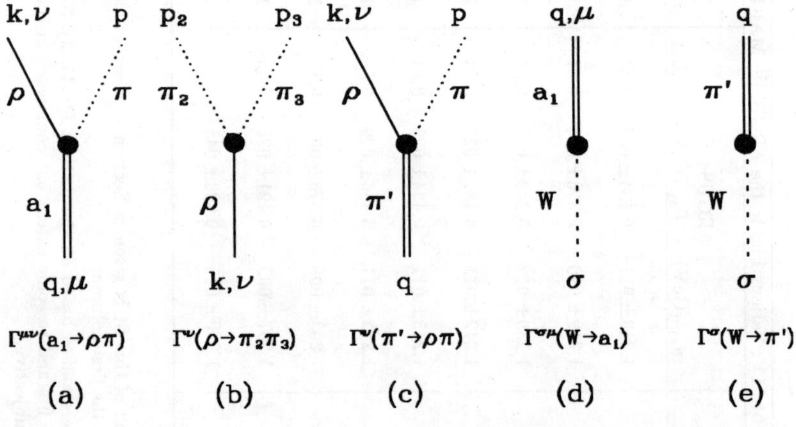

Figure 2: (a) the $a_1 \to \rho\pi$ vertex, (b) the $\rho \to \pi_2\pi_3$ vertex, (c) the $\pi' \to \rho\pi$ vertex, (d) the $W \leftrightarrow a_1$ vertex, (e) the $W \leftrightarrow \pi'$ vertex; in all cases the momentum flow of the lines is from the bottom to the top of the diagram so $q = k + p$ and $k = p_2 + p_3$.

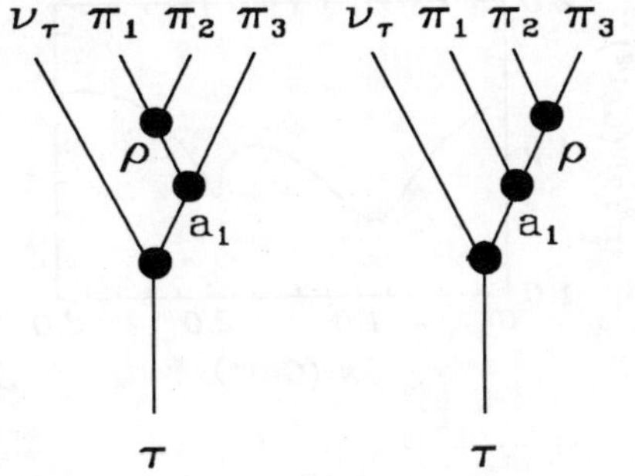

Figure 3: Covariant graphs contributing to $\tau \longrightarrow \nu_\tau \pi_1 \pi_2 \pi_3$ via the a_1 and ρ intermediate states. The blobs represent general vertex functions.

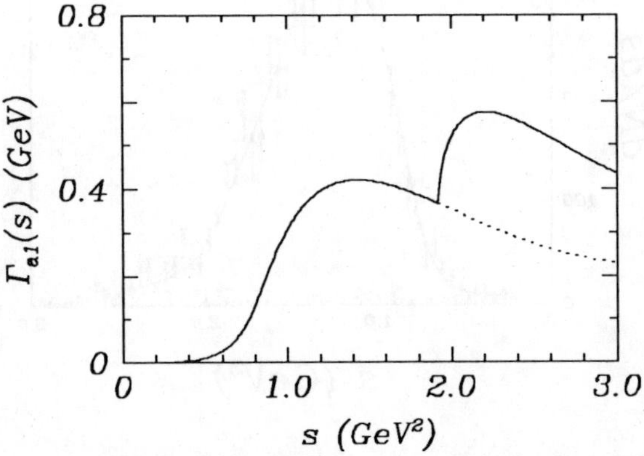

Figure 4: The total mass-dependent a_1 width (solid curve) from Eqn. (31) including the $\rho\pi$ and $K^*\bar{K}+\bar{K}^*K$ channels. The dotted curve indicates the $a_1 \rightarrow (\pi\pi)_\rho \pi$ partial width.

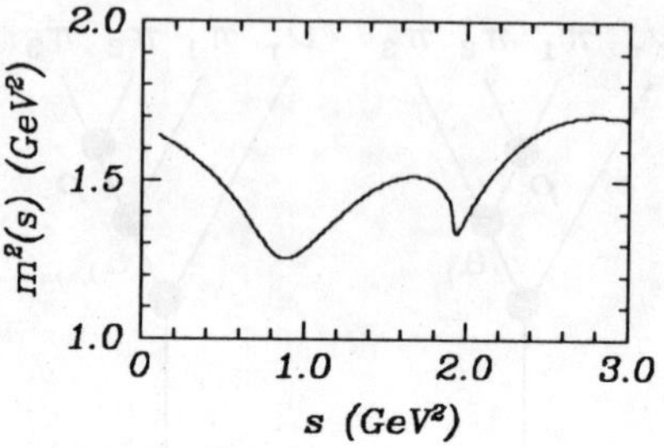

Figure 5: The a_1 mass (squared) function given by Eqn. (22).

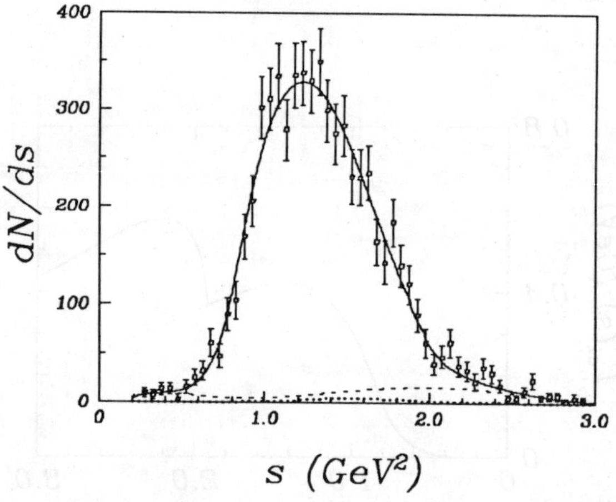

Figure 6: Fit to experimental three pion mass spectra[4] from $\tau \to \nu_\tau \pi\pi\pi$. The solid curve is the "preferred fit" described in the text. The dashed curve shows the fitted background polynomial and the dotted curve the predicted effect of the π'. The theoretical curve has been convoluted with a detector resolution function.

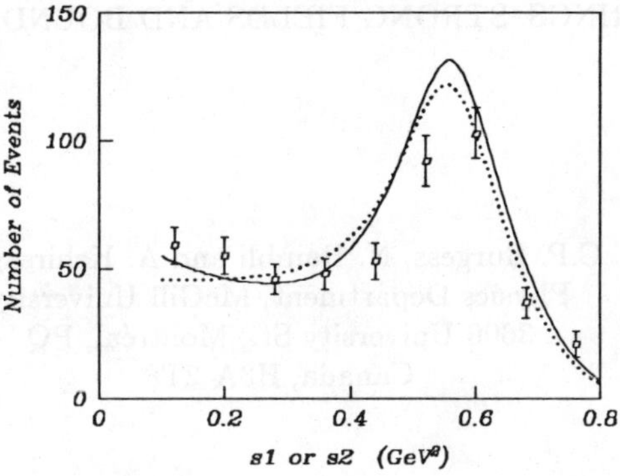

(a)

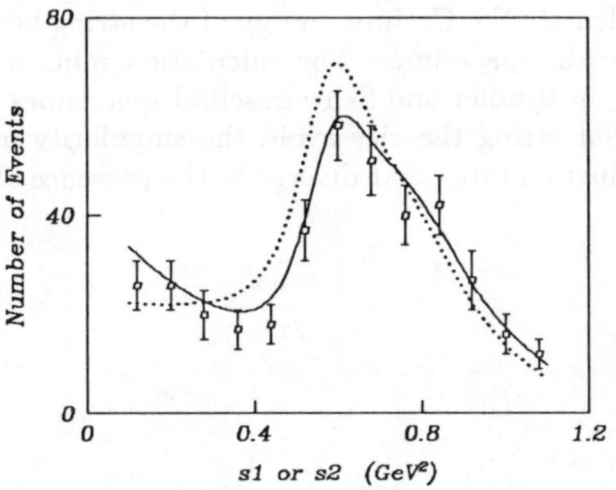

(b)

Figure 7: Comparison to experimental[4] two pion mass projections of the three pion Dalitz plot for various three pion mass bins. The solid curves correspond to the "preferred fits" described in the text and the dotted curves to the fits with the $a_1 \to \rho\pi$ D-wave amplitude arbitrarily set to zero (Fit 6 of Table IV). (a) 1.1025 GeV$^2 \leq m_{3\pi}^2 \leq$1.44 GeV^2, (b) 1.44 GeV$^2 \leq m_{3\pi}^2 \leq$1.96 GeV^2.

STRINGS, STRONG FIELDS AND BOUNDARIES*

C.P. Burgess, N. Hambli and A. Kshirsagar
Physics Department, McGill University
3600 University St., Montréal, PQ
Canada, H3A 2T8

ABSTRACT

String theory in spacetimes with boundaries is shown to be equivalent, in some instances, to string theory on orbifolds. This equivalence is used to calculate the Casimir energy of the string between two infinite parallel plates in spacetime. The calculation mimics some features of string theory in Rindler and Schwarzschild spacetimes and supports the conjecture that string theories avoid the singularity problem by giving particle-production rates that diverge in the presence of large, but finite, fields.

* Presented by C.B. to the M.R.S.T. meeting, Toronto Ontario, May 9-10, 1988.

1) INTRODUCTION:

At long last we have, in string theory, a candidate description of fundamental physics that appears to include gravity in a quantum-mechanically sensible way. It is difficult to overstate the significance of this achievement. Strings indeed are the *only* known framework in which this can be done. We are, at least in principle, in the happy position of finally being able to answer all of those questions whose understanding requires some knowlege of Planck-scale physics.

Despite this success a wide gap remains between our calculational aspirations and abilities, leaving many questions unanswered. We do not know whether string theory is consistent with what we see at low energies, largely because we are experimentally limited to energies many orders of magnitude lower than the fundamental string scale. We are also almost completely unable to calculate how string theories behave in the regime of very strong fields, where 'stringy' physics is expected to play an important, possibly dominant, role. In all of these examples we are handicapped by the absence of an efficient calculational scheme in which to properly pose these problems. Our present understanding of string theory relies on the few situations in which we have been able to calculate: *i.e.* in perturbation theory for weak fields and low energies. Any calculation that can broaden this base of experience on how string theory behaves in new physical situations is therefore of some interest.

The work described here is intended as a contribution along these lines. We analyze[1)] how strings behave in the presence of simple spacetime boundaries and apply this to compute the Casimir energy[2)] in string theory due to two infinite $(d-1)$-dimensional parallel plates. (d here is the dimension of space, 25 for the bosonic and 9 for the superstring, and the plates considered are $(d-1)$-dimensional is *space* but d-dimensional in *spacetime*.) Unlike the usual treatment of boundary problems for point-particle theories, our discussion is couched within the first-quantized Polyakov approach to string theory.

There are three main motivations for performing this calculation. The first of these, alluded to above, is the utility of any, preferably explicit, computation that extends our tiny beachhead of understanding of how strings behave in different physical situations.

The second motivation for this work is as a preliminary step towards an understanding of string behaviour in the presence of a black hole. It is well known that much of the physics of a point-particle theory in the presence of a black hole can be reproduced in the simpler problem of a mirror accelerating uniformly through flat spacetime[3]. This is because these two problems share many important features such as an event horizon and a set of preferred accelerating observers. The great advantage of the accelerating-mirror problem is that it is simple enough to be exactly soluble for point particle theories, and may be so for string theories too[4]. An obvious prerequisite for a string calculation with an accelerated mirror is some understanding of string theory in the presence of a stationary one.

The third motivation is perhaps the most interesting of the three. We know that at energies much less than the Planck scale the low-energy effective theory describing gravity is Einstein's theory of general relativity. This theory is a very good approximation for astrophysical systems and unambiguously predicts that large massive bodies ultimately collapse down to a singularity under their own gravitational attraction.[5] A similar prediction holds for the initial stages of the early universe. Once the spacetime curvature is of Planck size, the low-energy description breaks down and a more complete theory of Planck-scale physics is required in order to trace the further evolution of the collapsing body. Since string theory purports to be such a theory, one of its challenges is to provide some understanding of how these singularities are ultimately avoided.[6]

Most interestingly, there is a physical argument that, although imprecise, may indicate how strings can in general forbid the production of strong fields of any type. The Casimir energy is perhaps the simplest test case in which this argument can be compared to a direct calculation.

The argument proceeds as follows[7]. In general, even for point-particle theories, the presence of a strong field is associated with particle production from the vacuum. Since the produced particles carry away energy, they act to dissipate the applied field. The larger the field, the more particles are produced, and so the faster is the energy-dissipation rate. In string theories, as is argued below, at some finite critical field strength the particle-production rate, and hence the energy-dissipation rate, is expected to diverge. This implies that if energy is introduced in an effort to increase an applied field, more and more of this energy must be budgeted to replace the losses to radiation and less and less goes into strengthening the field. Ultimately, so the argument goes, at the critical field it becomes impossible to introduce enough energy to further increase the field. String theory would therefore automatically forbid the existence of regions of strong field and would thereby circumvent any singularities.

The key question is whether, in string theory, the particle-production rate really does diverge. The argument in favour of such a divergence is based on an analogy with similar reasoning that indicates the existence of a critical (Hagedorn) temperature.[8,9] Imagine estimating the string radiation rate by summing over the point-particle production rate for all of the string levels. Since particle production is essentially a tunnelling phenomenon in which the rest mass of the produced particles is extracted from the field, the rate for producing very massive string states is generically exponentially suppressed by terms involving the particle mass divided by the applied field strength. In some cases of interest, such as for Hawking radiation, the power of mass appearing in exponent of the rate is linear:

$$R(m,\varphi) \sim \exp\left(-A\frac{m}{\varphi}\right). \tag{1}$$

φ here is a measure of the strength of the applied field and 'A' is a dimensionless constant. The appoximate total rate for radiating strings found by summing over the rate for producing each string level then is:

$$R_s \approx \sum_m R(m,\varphi)\; N(m). \tag{2}$$

The sum is over the string levels of mass m with degeneracy $N(m)$. For large m $N(m)$ is given by:[9]

$$N(m) \sim \exp\left(B\frac{m}{\sqrt{T}}\right). \tag{3}$$

T here denotes the string tension and B is a known number. Comparison of eqs. (1) and (3) indicates that the sum in eq. (2) diverges for φ less than some critical value $\varphi_c \sim \sqrt{T}$.

Physically appealling as it may be, there are at least two things wrong with this argument:

(i) The string radiation rate is not necessarily correctly given by a sum over point-particle rates for each mass level. A string calculation effectively performs the sum over all string levels first and *then* integrates over the available momentum and energy. This differs from a collection of point-particle calculations in the order in which the sum and integration are done. Interchanging this order can lead to errors[10], particularly in situations for which the result is expected to diverge.

(ii) The exponential form, eq.(3), for the degeneracy, $N(m)$, is derived in flat, Euclidean 26- (resp. 10-) dimensional spacetime. Although this may be a reasonable approximation for field strengths much smaller than the string tension, this is bound to be suspect for φ near the critical value $\varphi_c \sim \sqrt{T}$.

It would therefore be instructive to test this reasoning in a regime for which a quantitative string calculation is possible. The Casimir energy due to two $(d-1)$-dimensional plane mirrors is just such a model system. The string Casimir energy per unit plate area, ϵ_s, due to the two mirrors a distance a apart can be estimated by summing the point-particle vacuum energy density, $\epsilon(m,a)$ over all string levels. Recall that in d space dimensions, and for $ma \gg 1$, the energy due to two plane $(d-1)$-dimensional plates is:

$$\epsilon(m,a) \sim p(m,a) \exp(-2ma) \tag{4}$$

where $p(m,a)$ is a monomial in m and a. ϵ_s might therefore be expected to be:

$$\epsilon_s(a) \sim \sum_m \epsilon(m,a)\, N(m). \tag{5}$$

This increases as the spacing, a, between the plates decreases and ultimately diverges when $a = a_c \sim 1/\sqrt{T}$. Notice that this argument exactly parallels the intuition described above concerning the string radiation rate. In this case, however, it can be compared to the full string calculation presented below to see whether it provides a reliable estimate or has led us astray. The conclusion reached below is that the one-loop string Casimir energy does indeed blow up at a critical plate separation, $a_c = \beta_H/2$, where β_H denotes the inverse of the Hagedorn temperaturefor the string. For the closed oriented bosonic string, $a_c = \sqrt{2\pi/T}$. This, in this case, vindicates the physical picture outlined above.

The calculation is presented in two parts below. The following section describes the first-quantized formulation of string propagation in the presence of spacetime boundaries and shows the relationship with orbifold calculations. For the purposes of illustration, the example of a single 24-dimensional plane mirror in the closed, oriented bosonic string is discussed in some detail. This is followed in section 3 by an application of this method to a computation of the Casimir energy due to two infinite parallel mirrors. Our conclusions are summarized in the final section.

2) FIRST-QUANTIZED STRINGS AND SPACETIME BOUNDARIES:

We here describe how to compute, within the Polyakov formulation, string behaviour in the presence of spacetime boundaries. The Polyakov approach to string calculations expresses stringy quantities, such as the S-matrix, in terms of functional integrals over Euclidean string world histories. This is the tool within which much, if not most, of what is presently known about string theory has been learned. It has the twin advantages of manifestly preserving many of the symmetries of the theory and being simple to compute with at low orders in string perturbation theory. Its principal disadvantage for the present purposes is that it is an awkward formalism in which to pose boundary-value problems. Most of the discussion in this chapter is therefore devoted to formulating boundary problems within such a first-quantized framework.

To do so it is easiest to consider how the problem would be handled in point-particle theories. Our strategy is to set the point-particle problem up using the more natural second-quantized field-theoretic language and then to translate the results into the comparatively cumbersome first-quantized form. Once expressed in this way it is straightforward to write down the generalization to first-quantized string theory.

(i) Point Particles and the Plane Mirror:

Consider for simplicity the theory of a single scalar field, $\psi(x^\mu)$, in $d+1$ spacetime dimensions, restricted to the half-space $x^1 \geq 0$, and satisfying Dirichlet ($\psi = 0$) or Neumann ($\partial\psi/\partial x^1 = 0$) at the boundary $x^1 = 0$. A discussion of the extent to which this analysis can be applied to a general system of boundaries is postponed to the end of this section. We are interested in computing the vacuum energy, E_v, in this theory due to the presence of the boundary. For time-independent field configurations this is related[11)] to the value of the Euclidean effective action, Γ, evaluated at its stationary point:

$$E_v \equiv \int d^d x \, \langle \mathrm{vac}|T^{00}(x)|\mathrm{vac}\rangle \\ = -\lim_{P\to\infty} \frac{\Gamma}{P} \tag{6}$$

The energy density, T^{00}, is the zero-zero component of the system's stress-tensor, $T^{\mu\nu}$. $P \equiv \int dx^0$ is the length of a fictitious 'box' in the Euclidean time direction.

Because of the symmetries of the problem E_v diverges like the volume of the directions of space parallel to the surface. The well-behaved quantity is the energy per unit area, denoted ϵ_v.

The bridge to the first-quantized formalism is made by relating Γ to the free propagator in the absence of a boundary, $G(x,x')$. A first-quantized path-integral representation for $G(x,x')$ is known:[12)]

$$G(x,x') = \int_x^{x'} \mathcal{D}x(s)\mathcal{D}e(s)\; e^{-S[x,e]}$$
$$\text{with:} \qquad S[x,e] = \frac{1}{2}\int_0^1 ds \left[\frac{m^2}{e(s)}\frac{dx^\mu}{ds}\frac{dx_\mu}{ds} + e(s)\right] \tag{7}$$

The functional integral is over all paths, $x^\mu(s)$, that run between the two given points x^μ and x'^μ in unit parameter distance. The functional integral over the 'einbein' along this path, $e(s)$, boils down to an ordinary integral over the proper time required to get from x to x'. m is the particle mass.

The connection between Γ and $G(x,x')$ is made via the Euclidean propagator, $\mathcal{G}_\pm(x,x')$, in the presence of the boundary:

$$(-\partial_\mu\partial^\mu + m^2)\mathcal{G}_\pm(x,x') = \delta^4(x-x'), \tag{8}$$

in terms of which the effective action is given by:

$$\Gamma = \frac{1}{2}\mathrm{Tr}\;\mathrm{Ln}\;\mathcal{G}_\pm$$
$$= \frac{1}{2}\int_{m^2}^{\infty} \mathrm{Tr}\;\mathcal{G}_\pm\; dm^2. \tag{9}$$

The subscript $\pm$ corresponds to the imposition of Dirichlet $(-)$ and Neumann $(+)$ boundary conditions, repectively, on the surface.

Finally, the relation between $\mathcal{G}_\pm(x,x')$ and $G(x,x')$ is given by the method of images:

$$\mathcal{G}_\pm(x,x') = G(x,x') \pm G(x,\tilde{x}') \tag{10}$$

in which $\tilde{x}^\mu = (x^0, -x^1, x^2, ..., x^d)$ is the 'image' point constructed from $x^\mu = (x^0, x^1, ..., x^d)$ by reflecting about the boundary at $x^1 = 0$.

Inserting the functional-integral representation, eq. (7), into eq. (10) and putting the result into the second of eqs. (9) gives a first-quantized expression for

Γ, and hence for ϵ_v. The functional trace has the effect of requiring functional integration over (based) loops that begin and end at a specific point x^μ or its mirror-image $\tilde{x}^\mu$. An integral over x^μ is performed at the end. Just as was the case in the absence of a mirror, performing the integration over m^2 converts $Tr\ \mathcal{G}_\pm$ into $Tr\ Ln\ \mathcal{G}_\pm$. This is equivalently accomplished[13] by functionally integrating over all unbased loops rather than over the based loops. The result is:

$$\Gamma = (\mathcal{Z}_+ \pm \mathcal{Z}_-) \tag{11a}$$

where:

$$\mathcal{Z}_\pm = \int_{x^1(s+1)=\pm x^1(s)} \mathcal{D}x(s)\mathcal{D}e(s)\ e^{-S[x,e]}. \tag{11b}$$

The $x^\mu(s)$ for $\mu \neq 1$ are strictly periodic in s. This is the first-quantized expression for the effective action and, by virtue of eq. (6), for the vacuum energy. The first term, $\mathcal{Z}_+$, represents the vacuum energy in the absence of any boundaries and so the second term $\mathcal{Z}_-$, gives the new boundary-dependent part of the energy.

Eqs. (11) have a suggestive form. They are exactly the functional integral describing the vacuum energy of a point particle moving on the orbifold,[14] $O = R^{d+1}/Z_2$, defined by identifying points in spacetime that are related by reflection through the plane $x^1 = 0$:

$$\Gamma = \mathcal{Z}[O] \tag{12}$$

Recall that the functional integral is over all unbased loops on O. These come in two topological classes and the contribution of each class must be summed in $Z[O]$. Eq. (11a) has exactly this form. The first term in this equation correponds to the integral over homotopically trivial closed loops while the second term is due to nontrivial loops that only close by virtue of the identification of points under reflection. For point-particle theories on an orbifold the relative sign of these two contributions can be chosen arbitrarily. In the operator interpretation of the path integral + (−) corresponds to a projection onto all particle states that are even (odd) under x^1-reflection. This freedom in orbifold theory corresponds to the choice of boundary conditions that must be made in the presence of the boundary in spacetime.

(ii) Strings and the Plane Mirror:

This way of writing the effective action generalizes directly to string theory. For simplicity we consider the closed, oriented bosonic string although the generalization to the superstring is straightforward. The string effective action, and hence vacuum energy, in the presence of a single infinite $(d-1)$-dimensional plane boundary is given by the Polyakov functional integral on the orbifold $O = R^{d+1}/Z_2$:

$$\Gamma = Z(++) + Z(+-) + Z(-+) + Z(--) \tag{13a}$$

where:

$$\begin{aligned} Z(\alpha\beta) &= \int_{(\alpha\beta)} \mathcal{D}g(\sigma)\mathcal{D}x(\sigma)e^{-S[x,g]} \\ \text{with:} \quad S[x,g] &= \frac{T}{2}\int d^2\sigma\sqrt{g}g^{\rho\sigma}\partial_\rho x^\mu \partial_\sigma x_\mu . \end{aligned} \tag{13b}$$

$(\alpha\beta)$ denotes the boundary conditions appropriate to maps of the torus into O:

$$\begin{aligned} x^\mu(\sigma^1+1,\sigma^2) &= \begin{cases} \alpha x^\mu(\sigma^1,\sigma^2), & \text{if } \mu = 1 \\ x^\mu(\sigma^1,\sigma^2), & \text{if } \mu = 0,2,\ldots,d \end{cases} \\ \text{and} \quad x^\mu(\sigma^1,\sigma^2+1) &= \begin{cases} \beta x^\mu(\sigma^1,\sigma^2), & \text{if } \mu = 1 \\ x^\mu(\sigma^1,\sigma^2), & \text{if } \mu = 0,2,\ldots,d \end{cases} \end{aligned} \tag{13c}$$

The first term, $Z(++)$, gives the effective action as would be computed in the absence of a boundary so the remaining terms may be interpreted as being the surface-dependent part of the vacuum energy. Denoting this latter surface-dependent vacuum energy per unit area by ϵ_s gives, after a simple calculation:[1)]

$$\epsilon_s = -T^{25/2}\int_F d^2\tau \frac{1}{2(2\pi\tau_2)^{27/2}}\left[\frac{|\vartheta_2\vartheta_3| + |\vartheta_2\vartheta_4| + |\vartheta_3\vartheta_4|}{|\eta|^{48}}\right] \tag{14}$$

in which $\vartheta_i = \vartheta_i(0|\tau)$ are the usual Jacobi ϑ-functions, $\eta(\tau) = [\vartheta_1'(0|\tau)/2\pi]^{1/3}$ is the Dedekind η-function and $\tau = \tau_1 + i\tau_2$ is the modular parameter of the torus to be integrated over the fundamental region: $F = \{|\tau| > 1, |\tau_1| < \frac{1}{2}, \tau_2 > 0\}$.

In the point-particle example, eq. (11a), the relative sign of the two contributions was at our disposal, corresponding to the choice of boundary conditions to be made at the surface. In string theory the analogous choice in eq. (13a) is fixed by general requirements of consistency. One such requirement is that the result be invariant under modular transformations of the torus:[15)]

$$\begin{aligned} (\sigma^1,\sigma^2) &\to (-\sigma^2,\sigma^1) \\ (\sigma^1,\sigma^2) &\to (\sigma^1-\sigma^2,\sigma^2) \end{aligned} \tag{15}$$

Modular invariance implies that $Z(+-)$, $Z(-+)$, and $Z(--)$ must appear in the effective action only in the combination $X = Z(+-) + Z(-+) + Z(--)$. Since the coefficient of $Z(++)$ in Γ is fixed by the method of images, the only remaining freedom is overall sign of X in eq. (13a). Unitarity requires that this sign be +, which is the choice appropriate to Neumann conditions in the point-particle analogue. Presumably a similar interpretation holds in terms of boundary conditions in a second-quantized formulation of string theory. One of the differences between the string and point-particle cases appears to be that in string theory these relative signs, and hence the boundary conditions, are fixed by general requirements such as modular invariance and unitarity.

Another feature of eq. (13a) not shared by the point-particle expression, eq. (11a), is the appearance of a 'twisted sector'. In the operator approach the first two terms of eq. (13a) correspond to a projection, within the usual closed-string Fock space, onto those states that are even under reflections in the surface $x^1 = 0$. The remaining two terms, however, describe the same projection in a string sector with twisted boundary conditions in the spatial direction perpendicular to the mirror: $x^1(\sigma^1 + 1, \sigma^2) = -x^1(\sigma^1, \sigma^2)$. The existence of this twisted sector is an intrinsically stringy phenomenon. It corresponds physically to string oscillations that are 'confined' to the surface of the mirror, in the sense that the string centre of mass in this sector never strays from $x^1 = 0$.

It is clear that this construction allows a wide class of point-particle boundary problems to be generalized into first-quantized string theory. The corresponding string problem can be constructed for any system to which the method of images can be applied in the point-particle case. The method of images, in turn, is applicable provided that the spacetime region of interest, B, (with boundary) can be described as the quotient of some spacetime, M, without boundary with respect to a group of symmetry transformations, G, of the theory. That is: $B = M/G$. The boundary itself is then the set of so-called 'orbifold' points left invariant under G. Furthermore, the boundary conditions chosen on this surface must be such that the eigenvalue problem on B can be formulated in terms of that on M by projecting onto a subspace of eigenfunctions that form an irreducible representation G. If these two conditions are satisfied then the point-particle effective action and vacuum energy on B can be written as a first-quantized functional integral over unbased loops on the orbifold $B = M/G$. The generalization to string theory is immediate provided that the background fields of the problem (*e.g.* the metric, dilaton, *etc.*) satisfy the classical string equations, as they must do to define a conformal field theory in the Polyakov approach. They are sure to do so on B provided that they do on M and are G-invariant.

3) THE CASIMIR ENERGY:

We now apply this reasoning to compute the Casimir energy of the closed, oriented bosonic string in the presence of two 24-dimensional infinite parallel plates. The motivation for this calculation is to test the physical reasoning outlined in the introduction. This reasoning doesn't strictly apply to the bosonic string since its vacuum energy diverges even in the absence of boundaries by virtue of the tachyon. We will therefore ignore the ordinary tachyon and search for any further divergence that appears only for plate separations, a, less than some critical value, a_c. The real motivation for this is that the bosonic string provides a stripped-down version of the superstring for which our conclusions[1] also hold but where the complications due to the tachyon do not arise.

Consider therefore string theory between two infinite plates described by the surfaces $x^1 = 0$ and $x^1 = a$. In keeping with our program to translate this into an orbifold calculation, we regard this region as being obtained from 26-dimensional flat space by identifying points under the group, G, generated by the following symmetry transformations of the closed-string action eq. (13b):

$$\begin{aligned} g: &\quad x^1 \to -x^1 \\ h: &\quad x^1 \to 2a - x^1 \end{aligned} \tag{16a}$$

corresponding to reflections in either of the two plates. An equivalent set of generators for this group is given by:

$$\begin{aligned} g: &\quad x^1 \to -x^1 \\ h': &\quad x^1 \to x^1 + 2a \end{aligned} \tag{16b}$$

Clearly $h = h'g$. Just as was the case for a single mirror, the point-particle vacuum energy can be computed here as a functional integral over all unbased loops on the quotient space, $B = R^{d+1}/G$. This space is the familiar orbifold[16] $B = R^d \times (S^1/Z_2)$, since identification under the shifts, h', compactifies the x^1-direction into a circle, S^1, of circumference $2a$, which is then identified under the reflections, Z_2, generated by g. The factor R^d describes the $d = 25$ directions parallel to the plates in space and time.

The one-loop string vacuum energy in the presence of the plates is therefore given by the functional integral over all torii embedded in the orbifold $B = R^d \times (S^1/Z_2)$. The toroidal string configurations on B can *a priori* be labelled by $(\alpha, \beta; m, n)$, where α and β are signs, $\pm$, and m and n are integers. The corresponding string configurations satisfy the boundary conditions:

$$x^\mu(\sigma^1+1,\sigma^2) = \begin{cases} \alpha x^\mu(\sigma^1,\sigma^2) + 2am, & \text{if } \mu = 1 \\ x^\mu(\sigma^1,\sigma^2), & \text{if } \mu = 0,2,...,d \end{cases}$$
$$\text{and} \quad x^\mu(\sigma^1,\sigma^2+1) = \begin{cases} \beta x^\mu(\sigma^1,\sigma^2) + 2an, & \text{if } \mu = 1 \\ x^\mu(\sigma^1,\sigma^2), & \text{if } \mu = 0,2,...,d \end{cases} \tag{17}$$

Not all of these boundary conditions are consistent. The consistent ones that give a distinct contribution to the functional integral are $(+,+;m,n)$, $(+,-;0,0)$, $(-,+;0,0)$ and $(-,-;0,0)$. The modular-invariant path integral describing the one-loop energy on B then is:[1,16]

$$\begin{aligned} &\Gamma[B] = \Gamma_0 + \Gamma_S + \Gamma_V(a) \\ \text{with:} \quad &\Gamma_0 = Z(+,+;0,0) \\ \text{and:} \quad &\Gamma_S = Z(+,-;0,0) + Z(-,+;0,0) + Z(-,-;0,0) \\ &\Gamma_V(a) = \sum_{mn}{}' Z(+,+;m,n) \end{aligned} \tag{18}$$

The first term, Γ_0, is the contribution expected in the absence of any boundaries. The second term, Γ_S, is the result from the remaining sectors with $m = n = 0$. The combination $\Gamma_0 + \Gamma_S$ is exactly the result computed in the previous section for the effective action in the presence of a single mirror. The prime on the sum in the final term, $\Gamma_V(a)$, indicates that the term with $m = n = 0$ is to be omitted. (Recall that the $m = n = 0$ piece is treated separately as Γ_0.) $\Gamma_V(a)$ and Γ_0 together make up the contribution from $R^d \times S^1$ (with no twists). $\Gamma_V(a)$ contains all of the dependence on the distance, a, between the plates and, as expected, vanishes in the limit that $a \to \infty$. Define the Casimir energy, ϵ_c, per unit area as the difference between the vacuum energy in the presence of, and absence of, the plates. Evaluating the path integrals[1,16] gives the following explicit form:

$$\epsilon_c = \epsilon_s + \epsilon_v(a) \tag{19}$$

where ϵ_s is the contribution due to a single plate, given in eq. (14), and the a-dependent part is:

$$\epsilon_v(a) = -2\pi a T^{13} \int_F d^2\tau \frac{1}{(2\pi\tau_2)^{14}} \frac{\Sigma(a,\tau)}{|\eta|^{48}}$$

$$\text{with:} \quad \Sigma(a,\tau) = {\sum_{mn}}' \exp\left(-\frac{2a^2T}{\tau_2}|m\tau - n|^2\right) \tag{20}$$

$$= \sqrt{\frac{\pi\tau_2}{2a^2T}} {\sum_{mn}}' \exp\left(-2a^2Tm^2\tau_2 - \frac{\pi^2 n^2 \tau_2}{2a^2T} - 2\pi i m n \tau_1\right).$$

This last form for the sum follows from the previous line by a Poisson resummation and is more useful for displaying the asymptotic form for large τ_2.

We are interested in whether or not ϵ_c develops a divergence for a less than some critical a_c. Inspection of its a-dependent part, eq. (20), shows that the sum on m and n is absolutely convergent so any divergence must occur in the integration over τ. As usual the only troublesome point is at the boundary of moduli space: $\tau_2 \to \infty$. The analysis from here on closely resembles the finite-temperature analysis.[17] The asymptotic form of the integrand of eq. (20) is:[18]

$$\epsilon_v(a) \approx -2\pi T^{25/2} \int_F d^2\tau \frac{1}{(2\pi\tau_2)^{27/2}} \Big[e^{(4\pi - 2a^2T)\tau_2} + e^{(4\pi - \pi^2/2a^2T)\tau_2} + \\ +96 e^{(2\pi - 2a^2T - \pi^2/2a^2T)\,\tau_2} \cos^2(2\pi\tau_1) + \ldots \Big] \tag{21}$$

The ellipsis denote terms of lower order in $e^{-\tau_2}$ together with terms that are of lower order once the τ_1 integral is performed. For large τ_2 the underlying torus is becoming degenerate and is dominated by the low-lying intermediate string states. The contribution from each state of mass M is proportional[10] to $\exp(-M^2\tau_2/2T)$. Each term in the asymptotic expansion, eq. (21), therefore corresponds to the contribution of a low-lying string level to the vacuum energy, and the spectrum can be read off from the argument of the exponential of Im τ. Terms that vanish due to the Re τ-integration correspond in the operator picture to states that are removed by the $L_0 = \overline{L}_0$ condition, which in this case is given in terms of the winding numbers, m and n, by $N - \overline{N} = mn$. For example, the first term in eq. (21) corresponds to the winding states with $n = \pm 1$ and $m = 0$ and no oscillators excited. Their mass is given by $M^2(0, \pm 1) = -8\pi T + 4a^2T^2$. Similarly, the second term corresponds to the first of the Kaluza-Klein excitations, $n = 0$ and $m = \pm 1$, for the ordinary closed-string tachyon with mass $M^2(\pm 1, 0) = -8\pi T + \pi^2/a^2$. This is followed by states with $mn = \pm 1$, one oscillator excitation and mass $M^2(\pm 1, \pm 1) = -4\pi T + 4a^2T^2 + \pi^2/a^2$ *etc.*

The contributions that are dangerous are the tachyonic ones for which $M^2 < 0$, since these give rise to an exponential divergence when integrating the imaginary part of τ. In keeping with our earlier remarks we will ignore the divergences due to the ordinary tachyon and its descendants with quantized Kaluza-Klein momentum in the x^1-direction. These correspond to states with $n = 0$ and no oscillator excitations but with arbitrary m. The motivation for this gross neglect is ultimately that the analogue of these states are absent in the superstring while the divergence that we will now identify persists even in the supersymmetric case.

Notice that all states with both m and n nonzero have $M^2 \geq 0$ so only the pure winding states, $n \neq 0$ with $m = 0$ and no oscillators, need be checked. Inspection of the above formula for the squared-mass of these states when $n = \pm 1$ shows that $M^2(0, \pm 1) = -8\pi T + 4a^2T^2 > 0$ for $a > a_c = \sqrt{2\pi/T} = \beta_H/2$. For smaller separations, $a < a_c$, the Casimir energy indeed diverges exponentially, just as would be expected from the estimate involving the sum over string states described in the introduction.

It is worth emphasizing that this constitutes a quantitative check on the intuition based on summing over string levels. The only approximation made is that the string coupling is small enough to justify computing the vacuum energy to lowest order in the loop expansion. In particular, nowhere is it assumed that the plate separation, a, be large compared to the string size, $T^{-1/2}$.

4) SUMMARY:

In this note we describe a formalism that allows the calculation of string properties in the presence of boundaries in real space (as opposed, for example, to world-sheet boundaries as encountered in open-string theories). The unusual feature of this approach is that it is couched in the first-quantized Polyakov formalism rather than in a second-quantized string field theory. This would be an extremely awkward way to compute for point-particle theories and the fact that it is simpler in string theory bears sad witness to the drawbacks inherent in present approaches to calculating in string field theory.

The method consists of translating the point-particle problem into a first quantized language. This is done by making use of the method of images to relate the point-particle propagator, and hence the vacuum energy, in the presence of a boundary to the propagator in the absence of boundaries. Expressing things in terms of this last propagator allows us to avail ourselves of its familiar first-quantized path integral representation. The vacuum energy ends up being expressed as a functional integral over unbased loops on an orbifold in which the singular, 'orbifold', points represent the boundaries themselves. This is a result that is easily generalized to string theory. The string vacuum energy is computed by evaluating the one-loop (genus one) string path integral on the corresponding orbifold.

There are several motivations for wishing to perform such an analysis, not the least of which is to examine how strings behave in a physical situation that has hitherto been inaccessible. An understanding of string propagation in the presence of stationary boundaries is also a prerequisite to a calculation of string behaviour in the presence of accelerating boundaries. This latter calculation is of some interest as an analogue for the more difficult problem of string propagation in the presence of black holes.

Ultimately the goal of this line of investigation is to learn something about how strings behave in the presence of strong fields, such as those arising due to gravitational collapse. There are arguments that appear to indicate that string theory can forbid the appearance of ultra-strong fields by converting coherent field energy into incoherent radiation energy much more efficiently than do point-particle theories. If this way of thinking about string theory is correct, then the calculation of the total string radiation rate should behave similarly to the string partition function at finite temperature. That is, if the string rate is calculated by summing over the rate for producing each of its levels, the exponentially large degeneracy of these levels overwhelms the exponentially small rate to produce any given particle and gives a divergent energy dissipation rate in the presence of a strong but finite-strength field. This would furnish an appealling physical picture in which all of the energy intended to be used to produce a strong field in string theory gets persistently misdirected into producing radiation.

Appealling though this picture may be, it is based on several shaky assumptions. In particular, it need not be true that point-particle production rates summed over the string spectrum reproduce a full string calculation. Short of a full calculation of string behaviour in the presence of strong fields, what is necessary is a 'theoretical laboratory' in which the intuition furnished by summing over the contributions of string levels can be confronted with a quantitative string calculation. The Casimir energy in string theory between two plates is just such a testing ground. If summed over string levels, the intuitive argument predicts that the string Casimir energy should diverge once the plate separation comes below some critical value, a_c. The full one-loop string calculation presented here reproduces this behaviour and indeed blows up for $a < a_c = \beta_H/2$ ($= \sqrt{2\pi/T}$ for the closed, oriented bosonic string). It does so because the translation of the boundary problem to an orbifold problem essentially reduces the calculation to the behaviour of a string in spacetime with one dimension compactified to a circle. As is well known, such a calculation is closely related to the string partition function and so the divergence found exactly corresponds to that associated with the Hagedorn temperature. This is, after all, exactly the problem on which our string intuition is ultimately based.

Confirming, as it does in this instance, the intuitive understanding of string behaviour, the present calculation gives further motivation to improving our understanding of string physics in the presence of strong fields.

ACKNOWLEDGEMENTS

We would like to acknowledge useful conversations with Louise Dolan, and to thank the organizers of the Montreal-Rochester-Syracuse-Toronto meeting for the opportunity to present our work. This research is supported in part by NSERC.

bigskip

References

1) C.P. Burgess, N. Hambli, and A. Kshirsagar, McGill University preprint, July 1988.
2) H.B.G. Casimir, Proc. Kon. Ned. Akad. Wet. **51** (1948) 793.
3) N.D. Birrel and P.C.W. Davies, *Quantum Fields in Curved Space*, (Cambridge University Press, 1982) and references therein.

4) N. Sakai, Tokyo preprint (1986);
N. Sanchez and H.J. de Vega, CERN preprints (1987).
5) S.W. Hawking and G.F.R. Ellis, *The Large-Scale Structure of Space-Time*, (Cambridge University Press, 1973).
6) String quantum effects in the presence of strong fields and gravitational collapse have been considered in different contexts by:
M.J. Bowick, L. Smolin, and L.C.R. Wijewardhana, Phys. Rev. Lett. **56** (1986) 424; Gen. Rel. Grav. **19** (1987) 113;
I. Antoniadis, G.F.R. Ellis, J. Ellis, C. Kounnas and D.V. Nanopoulos, CERN preprint (1987);
N. Sanchez and H.J. de Vega, CERN preprints (1987).
7) N. Sakai, Tokyo preprint (1986);
C.P. Burgess, Nucl. Phys. **B294** (1987) 427.
8) R. Hagedorn, Nuovo Cimento Suppl. **3** (1965) 147.
9) K. Huang and S. Weinberg, Phys. Rev. Lett. **25** (1970) 895;
S. Fubini and G. Veneziano, Nuovo Cimento **64A** (1969) 1640.
10) J. Polchinski, Commun. Math. Phys. **104** (1986) 37.
11) S. Coleman, *Laws of Hadronic Matter*, edited by A. Zichichi (periodici Scientifici, Milan, 1975);
T.D. Lee and G.C. Wick, Phys. Rev. **D9** (1974) 2291.
12) R.P. Feynman and A.R. Hibbs, *Quantum Mechanics and Path Integrals*, (McGraw-Hill, New York 1965);
G. Moore and P. Nelson, Nucl. Phys. **B266** (1986) 58.
13) Technically the difference is due to the 'conformal killing vector' that must be omitted in the integral over $e(s)$. This vector only satisfies the boundary conditions appropriate to unbased loops. See for example:
I Giannakis, C. Ordóñez, M. Rubin and R. Zucchini, Rockefeller preprint RU88/B1/28 (1988).
14) L. Dixon, J. Harvey, C. Vafa and E. Witten, Nucl. Phys. **B261** (1985) 678.
15) N. Seiberg and E. Witten, Princeton preprint (1986).
16) P. Ginsparg, Harvard preprint HUTP-87/A068 (1987);
J. Bagger, Harvard preprint HUTP-87/A074 (1987); proceedings of *the 11th Johns Hopkins Workshop on Current Problems in Particle Theory*, (Lanzhou, China, 1987).

17) D. Gross, J. Harvey, E. Martinec and R. Rohm, Nucl. Phys. **B256** (1985) 253;
E. Alvarez, Phys. Rev. **D31** (1985) 418;
B. Sundborg, Nucl. Phys. **B254** (1985) 583;
M.J. Bowick and L.C.R. Wijewardhana, Phys. Rev. Lett. **54** (1985) 2485;
S.H. Tye, Phys. Lett. **158B** (1985) 388;
B. McClain and B. Roth, Commun. Math. Phys. **111** (1987) 539;
B. Sathiapalan, Phys. Rev. **D35** (1987) 3277;
A. Kogan, ITEP preprint (1987).

18) C.P. Burgess, Nucl. Phys. **B284** (1987) 605 (appendix).

TOPOLOGICAL OBJECT IN QUANTUM GRAVITY

C. Aneziris, A. P. Balachandran, M. Bourdeau, S. Jo and R. Sorkin
Physics Department
Syracuse University
Syracuse, N.Y. 13244-1130

T. R. Ramadas
School of Mathematics
The Institute for Advanced Study
Princeton, N.J. 08540

ABSTRACT

We discuss the spin and statistics of particle-like topological objects, called geons, in quantum gravity. It is shown that different spins and statistics can arise through quantization. We also show that certain geons can be quantized so that they are characterized by no definite statistics.

I. INTRODUCTION

A generally covariant theory is characterized by the diffeomorphism invariance which is related to the general coordinate transformation invariance. In this theory there are a class of particle-like topological excitations called "geons" discovered by Friedman and Sorkin and discussed by these authors and Witt.[1,2,3] We will begin with a brief explanation of geons.

In the following we will consider generally covariant theories built on asymptotically flat spaces M assuming that there are no matter fields present for simplicity. Then the configuration space Q is G/D^* where G is the collection of asymtotically flat metrics on a given space M and D^* is the group of diffeomorphisms of M leaving infinity and a frame at infinity invariant. A point $[g]$ in Q is an equivalent class of metrics obtained by acting D^* on g. Corresponding to the point $[g]$ we can have an imbedding of the space

M in a sufficiently higher dimensional Euclidean space such that the metric induced on this imbedded space from the flat Euclidean metric coincides with the original metric g. [Note that diffeomorphisms (D^*) do not change the imbedding.] Different points in Q will lead to different imbeddings. The dynamics in the second picture will show up as the motion of the imbedded surface while in the first picture the metric will change as time varies. Furthermore for a given d dimensional space M we can always decompose this into a connected sum of elementary building blocks, prime summands, which we explain now. The connected sum of two given d dimensional manifolds M and N, $M\#N$, is defined as follows. We first remove a d dimensional ball from each of the two and connect the two along the boundaries $(d-1)$ spheres arising from the removal of the balls. A manifold M is decomposable if it can be written as a connected sum of two manifolds both of which are topologically differernt from the original. If a manifold is not decomposable, we call it a prime summand. Note that a d sphere S^d is not considered as a prime summand because for any d dimensional manifold the following is true:

$$M\#S^d = M$$

A space with one asymtotic region which we will consider here can always be thought of as the connected sum of R^d and a finite number of compact prime summands and these compact prime summands are called geons. For $d = 2$ there are two prime summands, the torus T^2 and real projective sphere RP^2, while for $d = 3$ there are infinitely many prime summands and they have been partially classified. A real projective sphere RP^d can be obtained by identifying opposite points of a d sphere S^d.

Combining this decomposition with our second picture of imbeddings we see that the dynamics will show up as the motion of geons (compact prime summands) on an asymptotic flat surface R^d. Even though it sounds interesting to investigate the dynamical aspect of a given theory, we will not dwell on this problem. Instead we will consider only those aspects which come from the topology of the given manifold, such as spin and statistics for quantized geons.

In the following section we briefly review how various spins and statistics arise in quantum mechanics. In section III we discuss the spin and statistics of geons in quantum gravity. We show that the spin of the T^2 geon can be fractional while RP^2 and RP^3 geons can have only zero spin. We also show that the RP^2 geon can be characterized by θ statistics or no definite statistics at all and similarly the RP^3 geon can be characterized by either bosonic or fermionic statistics or by no definite statistics at all.

II. SPIN AND STATISTICS IN QUANTUM MECHANICS

In quantum theory wave functions are not functions on the configuration space Q but on the universal covering space $\tilde{Q}$ which is the principal bundle with the base manifold Q and the structure group $\Pi_1(Q)$, the fundamental group of Q. However wave functions should represent the group, $\Pi_1(Q)$, unitarily. As was discussed by Balachandran[4] there are as many ways of quantizing a classical system with the configuration space Q as there are unitary irreducible representations (UIR's) of the fundamental group, $\Pi_1(Q)$. An element of the group as a nontrivial loop in Q corresponds to a physical action such as a rotation or a permutation etc. By looking at the reaction of a wave function under the action of this element, for example rotation or permutation, we learn the information about the spin or the statistics of the particle described by this wave function. To be more specific we first consider spin and then statistics.

II.a Spin

In order to allow nontrivial spin we assume that a particle in R^d has a finite volume with an arbitrary but rigid shape. The configuration space Q of this system is $R^d \times SO(d)$, which is the frame bundle of R^d. The fundamental group, $\Pi_1(Q)$, is known for any d:

$$\Pi_1(Q) = Z_2, \quad d \geq 3$$

$$\Pi_1(Q) = Z, \quad d = 2$$

The generating element of the fundamental group is the loop in Q corresponding to a 2π rotation. For $d \geq 3$ the group is Z_2 and there are two inequivalent UIR's of Z_2, the trivial one, in which the wave function is invariant under the rotation, and the nontrivial one, in which the wave function changes sign under the rotation.

$$R(2\pi)\Psi_1 = \Psi_1, \quad \text{for the trivial UIR}$$

$$R(2\pi)\Psi_2 = -\Psi_2, \quad \text{for the nontrivial UIR}$$

Because spin S_n of the particle described by a wave function is defined by

$$R(\theta, \mathbf{n})\Psi(S_n) = \exp(iS_n\theta)\Psi(S_n),$$

it follows that the trivial UIR describes a boson and the nontrivial one describes a fermion. Therefore if $d \geq 3$ there are only two possibilities, boson or fermion. However if $d = 2$ the fundamental group is Z and its UIR is characterized by an angle θ,

$$R(2\pi)\Psi = e^{i\theta}\,\Psi,$$

describing fractional spin particles. The examples of the fractional spin have been studied by several authors.[5,6]

II.b Statistics

Consider a system with N identical particles in R^d. The configuration space Q for N identical particles has coordinates $[x^{(1)}, x^{(2)}, \cdots, x^{(N)}]$ where $x^{(i)} \in R^d$ and where we identify $[x^{(1)}, x^{(2)}, \cdots, x^{(i)}, \cdots, x^{(j)}, \cdots, x^{(N)}]$ with $[x^{(1)}, x^{(2)}, \cdots, x^{(j)}, \cdots, x^{(i)}, \cdots, x^{(N)}]$ in view of the identity of the particles. For technical reasons, the particles are also forbidden to occupy the same position so that $x^{(i)} \neq x^{(j)}$ for $i \neq j$. The fundamental group of this Q is known to be[7]

$$\Pi_1(Q) = S_N \qquad \text{for} \quad d \geq 3$$

$$\Pi_1(Q) = B_N \qquad \text{for} \quad d = 2.$$

When $d \geq 3$, it is the permutation group S_N which defines the usual statistics. However when $d = 2$, it is an infinite discrete group called the braid group B_N. The group B_N is generated by $N-1$ elements $\{a_1, a_2, \cdots, a_{N-1}\}$ with the following admissible presentation:

$$a_i a_{i+1} a_i = a_{i+1} a_i a_{i+1}, \qquad i = 1, 2, \cdots, N-2$$

Note that each element of $\Pi_1(Q)$ corresponds to a change of the location of the particles. From the above we can read that for $d \geq 3$ the path of the exchange is not important while for $d = 2$ it is important. Now let us consider a two particle system. When $d \geq 3$, the fundamental group is S_2 which is isomorphic to Z_2 and its UIR is either the trivial one describing bosonic statistics or the nontrivial one describing fermionic statistics. However when $d = 2$, the group B_2 is isomorphic to Z and its UIR is again chracterized by an angle variable θ, describing θ statistics.[6,7,8]

III. SPIN AND STATISTICS IN QUANTUM GRAVITY

The argument of the previous section can be summarized as follows. First, find the fundamental group of the configuration space for a given system. Second, find the all possible UIR's of the fundamental group. Third, find the element of the group corresponding to a rotation or an exchange. Finally see how this element is represented by each UIR and this will determine the spin or the statistics. We first consider the spin of geons of T^2, RP^2 and RP^3. Then we consider the statistics of geons of RP^2 and RP^3. In all the cases we consider in this section the fundamental group of the configuration space $Q\,(= G/D^*)$ is D^*/D_0^* where D_0^* is the connected component to the identity of D^* because the space G is simply connected. In this note we shall skip the details of the derivation of the fundamental group which will appear elsewhere.

III.a Spin of Geons

In order to find the possible spin of the individual geon we need to consider only one geon system. Let us first consider a T^2 geon. The space we treat is $M = R^2 \# T^2$. The fundamental group $\Pi_1(Q) = D^*/D_0^*$ is generated by two Dehn twists, h_a and h_b, along two nontrivial simple (without any crossing) closed loops a and b which form a complete homology basis for M. [A Dhen twist for a given simple closed loop is defined by first cutting along the loop and second turning one side counterclockwise by 2π along the other side while keeping the other side fixed and finally attaching them together.] The group has the following presentation:

$$h_a h_b h_a = h_b h_a h_b$$

Note that this group is isomorphic to B_3 discussed in the previous section. Even though we were not able to classify all the UIR's of this group, we do not need all of them if we are interested only in the spin of the geon. The element corresponding to the rotation is $(h_a h_b h_a)^4$ which is in the center of the group. Therefore we have only to consider the $U(1)$ representation of the group. Because $U(1)$ is an Abelian group, using the relation (III.1) we see that h_a and h_b are represented by the same $U(1)$ element, say $e^{i\theta}$, which in fact defines the $U(1)$ representation. In this representation the wave function develops phase factor $e^{12i\theta}$ under the rotation and describes a fractionally spinned geon.

For one RP^2 geon, $M = R^2\#RP^2$, and one RP^3 geon, $M = R^3\#RP^3$, the fundamental group turns out to be trivial,[3] hence allowing no nonvanishing spins for these geons.

III.b Statistics of Geons

We first consider two RP^2 geons. The space we treat is $M = R^2\#RP^2\#RP^2$. The corresponding fundamental group is $Z \times Z$, where $\times$ means a semidirect product. This group is generated by two elements p and d satisfying

$$dp = pd^{-1}.$$

The element p corresponds to a counterclockwise π rotation, hence changing the location of two RP^2's, while the other element d is a Dehn twist along an orientation preserving simple closed loop which passes through each RP^2 once. The UIR's of this group are classified by two real angle variables θ and ϕ. When $\phi = 0$, we get a one dimensional representation,

$$R_{\theta 0}(p) = e^{i\theta}, \quad R_{\theta 0}(d) = 1$$

and when $\phi = \pi$, we get another one dimensional representation,

$$R_{\theta\pi}(p) = e^{i\theta}, \quad R_{\theta\pi}(d) = -1.$$

For the other values of ϕ we get two dimensional representations:

$$R_{\theta\phi}(p) = e^{i\theta}\begin{pmatrix} 0 & 1 \\ 1 & 0 \end{pmatrix}, \quad R_{\theta\phi}(d) = \begin{pmatrix} e^{i\phi} & 0 \\ 0 & e^{-i\phi} \end{pmatrix}$$

As was mentioned the element of our interest is p corresponding to the exchange of two RP^2's. This element is represented by $e^{i\theta}$ in both of the one dimensional representations. Therefore we see again the occurrence of the θ statistics which we have seen in two dimensional quantum mechnics. However in two dimensional representations this element is represented by $R_{\theta\phi}(p)$ as a two by two matrix. [Note that in high dimensional representations wave functions are not just functions but vector valued functions.] The eigenvalues of $R_{\theta\phi}$ are $\pm e^{i\theta}$, indicating that the statistics of the corresponding eigenstates again describe θ statistics geons. However for generic states which are not eigenstates we have no definite statistics.

We now consider a three dimensional example, two RP^3 geons. The space for two RP^3 geons is $M = R^3\#RP^3\#RP^3$. The fundamental group of the configuration space is generated by two elements p and s with the following relation

$$p^2 = s^2 = 1.$$

The first element p is the exchange of the two RP^3's while the other element s corresponds to first sliding one RP^3 through the other RP^3 and then returning it back to the original position. [To have a better picture of s assume that before sliding we reduce the size of one RP^3 until it becomes very small compared to the size of the other RP^3 and after sliding the former through the latter we enlarge the former to the origonal size.] There are four inequivalent one dimensional representations:

$$R_1(p) = 1, \qquad R_1(s) = 1$$

$$R_2(p) = 1, \qquad R_2(s) = -1$$

$$R_3(p) = -1, \quad R_3(s) = 1$$

$$R_4(p) = -1, \quad R_4(s) = -1$$

The first two representations R_1 and R_2 describe bosonic statistics while the other two representations R_3 and R_4 describe fermionic statistics. As in the case of two RP^2's we also have two dimensional UIR's. These are classified by one parameter β with $-1 < \beta < 1$,

$$R_\beta(p) = \begin{pmatrix} 1 & 0 \\ 0 & -1 \end{pmatrix}, \quad R_\beta(s) = \begin{pmatrix} \beta & \gamma \\ \gamma & -\beta \end{pmatrix}$$

where $\gamma = \sqrt{1-\beta^2}$. Therefore we have another example of indefinite statistics.

Finally let us mention that, as Sorkin emphasized elsewhere,[1] there are no correlations between spin and statistics. We previously showed that the intrinsic spin of RP^2 or RP^3 is zero while the possible statistics for each of them include both bosonic and fermionic statistics.

The work of C. A., A. P. B., M. B. and S. J. is supported by DOE under contract number DE-FG02-85ER40231. The work of T. R. R., and R. S. is supported by NSF under contract numbers DMS 8601798 and PHY 8700651 respectively.

References

1. R. Sorkin in *Topological Properties and Global Structure of Space-Time*, ed. P. G. Berbmann and V. D. Sabbata (Plenum Publishing Corporation, 1986).
2. J. L. Friedman and D. M. Witt, Phys. Lett. **B120**,324 (1983).
3. D. M. Witt, J. Math. Phys. **27**, 573 (1986).
4. A. P. Balachandran, preprint, SU-4428-361 (1987).
5. F. Wilczek, Phys. Rev. Lett. **48**, 1144 (1982).
6. F. Wilczek and A. Zee, Phys. Rev. Lett. **51**, 2250 (1983).
7. Y.-S. Wu, Phys. Rev. Lett. **52**, 2103 (1984).
8. Y.-S. Wu, Phys. Rev. Lett. **53**, 111 (1984).

TWO GENERAL ASSERTIONS IN PERTURBATIVE QCD AND SUPERSYMMETRIC QCD*

A.P. Contogouris, N. Mebarki and H. Tanaka
Department of Physics, McGill University, Montreal, Canada

ABSTRACT

First, the following general assertion in perturbative QCD, and in Supersymmetric QCD at ultrahigh energies, is demonstrated and discussed: For processes involving structure functions and/or fragmentation functions, as p_T increases for fixed s, soft and collinear gluons dominate the Bremsstrahlung contributions; and their dominance increases with the softness of the structure and fragmentation functions. A way to determine the dominant part (without recourse to the complete Bremsstrahlung calculation) in a specific example is also presented.

Second, the following general assertion regarding changes of the renormalization and factorization scales (μ and M) is discussed: Consider a complete calculation of higher order corrections (HOC) and the corresponding overall inclusive cross section (Born + HOC) $\sigma = \sigma(\mu, M)$. Suppose an approximation of σ (i.e. an approximation of HOC) is good at the physical scale, say $\mu = M = p_T$; then it remains a good approximation for wide variations of μ, M away from p_T. This implies very similar results regarding optimization procedures, the same stability against changes of μ and M, etc.

* Also supported by the Natural Sciences and Engineering Research Council of Canada and by the Quebec Department of Education.

THE FIRST GENERAL ASSERTION

1. INTRODUCTION

In perturbative QCD, higher order corrections (HOC) have been calculated for many processes. In particular when the Born term is of order α_s or higher, the calculations are very involved and the expressions very complicated. Yet, almost invariably, the result is very simple: an overall inclusive cross section differing from the Born by a slowly varying factor.(1)(2)

This suggests that there is a relatively simple dominant part of the HOC which, if identified, can perhaps be calculated easier. If so, this could be useful in various directions, as e.g. in determining approximate HOC for QCD subprocesses of the type $a+b \to c+d$ and particularly $a+b \to c+d+e$ for which complete HOC are hitherto unknown (due to their complexity), and in going beyond the next-to-leading order in α_s.

Here we show that for processes involving structure functions and/or fragmentation functions there is indeed such a part which dominates HOC over a large kinematic domain. We show that the Bremsstrahlung (Brems) contributions to this part correspond to emission of soft and collinear gluons and that this part can be determined much easier.

2. THE STRUCTURE OF HIGHER ORDER CORRECTIONS AND THE DOMINANT PART

We consider large-p_T direct photon production and in particular the difference of inclusive cross sections

$$\sigma \equiv E\frac{d\sigma}{d^3p}(\bar{p}p \to \gamma X) - E\frac{d\sigma}{d^3p}(pp \to \gamma X) \tag{2.1}$$

which is dominated by the subprocess $q\overline{q} \to \gamma g$. This can be written:

$$\sigma = \frac{\alpha_s(p_T)}{\pi} \int dx_a dx_b F_{a/A}(x_a, p_T) F_{b/B}(x_b, p_T) \{ B\delta(1 + \frac{\hat{t}+\hat{u}}{\hat{s}})$$

$$+ \frac{\alpha_s(p_T)}{\pi} \cdot f \cdot \theta(1 + \frac{\hat{t}+\hat{u}}{\hat{s}}) \} \tag{2.2}$$

where $F_{a/A}, F_{b/B}$ are structure functions (here of q and $\overline{q}$), B is the Born term and f stands for the HOC.

We introduce the following dimensionless variables

$$v \equiv 1 + \frac{\hat{t}}{\hat{s}} \qquad w \equiv -\frac{\hat{u}}{\hat{t}+\hat{s}} \tag{2.3}$$

so that

$$\hat{s} + \hat{t} + \hat{u} = \hat{s}v(1-w)$$

Thus, considering Fig. 1 ($x_T = 2p_T/\sqrt{s}$, rapidity $y = 0$), in Eq. (2.2) the Born term contributes on the boundary $w = 1$ and the HOC in the hatched region $w \leq 1$.

Now it follows from a number of explicit results[(1),(3)] and will also become clear in the next section that the general structure of HOC is as follows:

$$f(v,w) = f_s(v,w) + \tilde{f}(v,w) \tag{2.4a}$$

where

$$f_s(v,w) = a(v)\delta(1-w) + b(v)\frac{1}{(1-w)_+} + c(v)(\frac{\ell n(1-w)}{1-w})_+ \tag{2.4b}$$

The function $\tilde{f}(v,w)$ is smooth as $w \to 1$ and, in general, very complicated (hundreds of terms); it is the most complicated part of HOC.[(1),(3)] Notice that each of the parts of f_s is gauge invariant.

It will become clear that a, b and c receive contributions from soft and collinear gluon Brems (e.g. Fig. 2(a), (b)); $\tilde{f}(v,w)$ receives contributions from hard and non-collinear gluon Brems. Graphs corresponding to $q\overline{q}$ production (Fig. 2(c))contribute

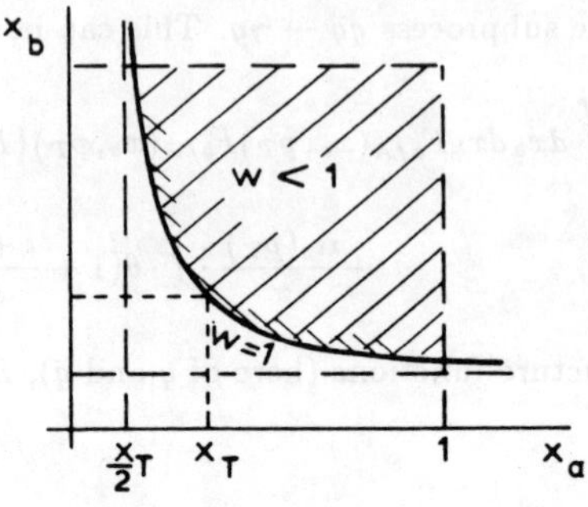

Fig. 1

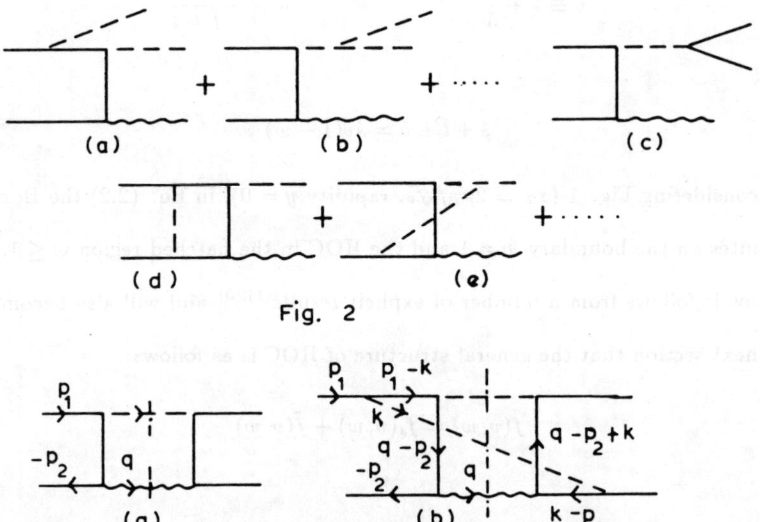

Fig. 2

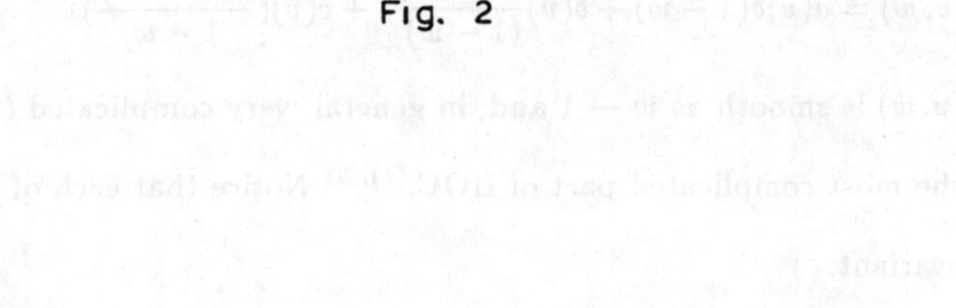

Fig. 2′

to both $\tilde{f}$ and f_s. Finally, a receives also contributions from loop graphs involving virtual gluons (Fig. 2(d), (e)).

Now denote by σ_s the contribution to the inclusive cross section σ of the part f_s of (2.4a) and by σ_{HO} that of the complete f; thus $\sigma_{HO} - \sigma_s$ corresponds to the contribution of $\tilde{f}$. Consider the ratio

$$\frac{\sigma_{HO} - \sigma_s}{\sigma_{Born} + \sigma_{HO}}; \tag{2.5}$$

Fig. 3 (solid line, denoted by $n = n_{QCD}(p_T)^{(4)}$) shows that, considered as function of p_T for fixed s, this ratio is small and decreases with p_T.

To understand the reason, a crucial observation is that $F_{a/A}(x, p_T)$ behaves like $(1-x)^n$; with A =proton, n is quite large ($n \sim 3$ for u-quark, $n \sim 4$ for d, $n \sim 5$ for gluon). Notice also that the scale violations enhance n as p_T increases. The same for $F_{b/B}(x_b, p_T)$. Then, referring to Fig. 1, contributions arising from the region away from $w = 1$ are suppressed by powers of $1 - x_a$ and/or $1 - x_b$. Now, the terms of f_s of (2.4) contribute at $w = 1$ or mainly at $w \simeq 1$ (cross hatched region of Fig. 1). However, the multitude of terms of $\tilde{f}$ do not mainly contribute at $w \sim 1$ and they are suppressed. As p_T increases for fixed s (i.e. as x_T increases) the boundary of the hatched region (Fig. 1) moves towards $x_a = x_b = 1$ and the whole region shrinks; thus the suppression increases.

To test our ideas, we have carried the following calculations: Writing the structure functions in the form

$$F_{a/A}(x, p_T) = F_{b/B}(x, p_T) = (1-x)^n \tag{2.6}$$

we have calculated $(\sigma_{HO} - \sigma_s)/(\sigma_{Born} + \sigma_{HO})$ for the fictitious values $n = 20$ (extremely soft distribution) and $n = 0.01$ (extremely hard). As expected, in the first case (dashed

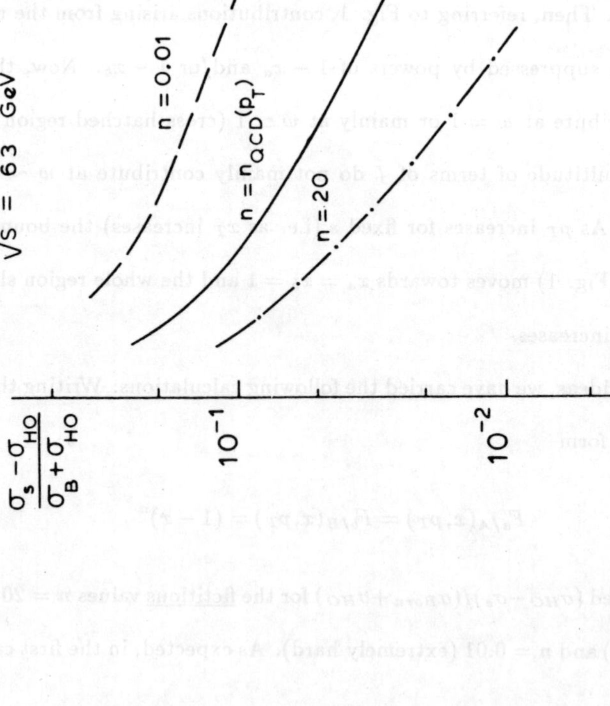

qq̄ → γg
√S = 63 GeV
(σ_s − σ_HO)/(σ_B + σ_HO)
10^-1
10^-2
n = 0.01
n = n_QCD (p_T)
n = 20
4
8
12
20
p_T

Fig. 3

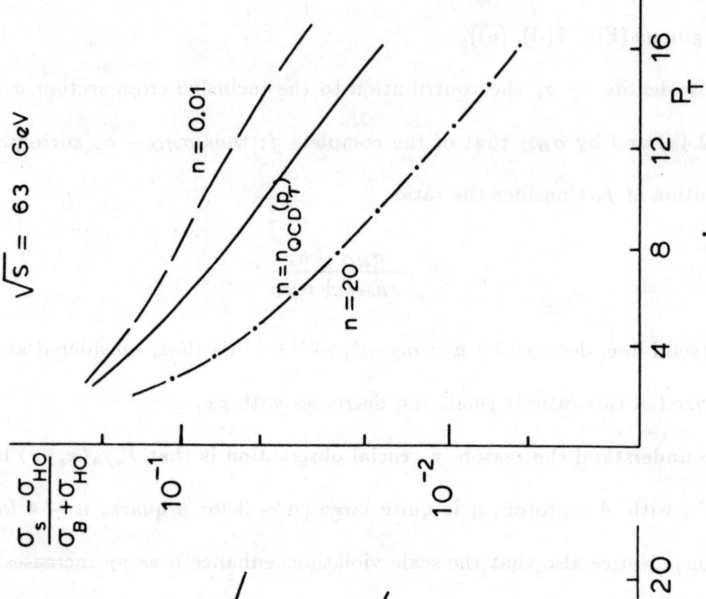

qg → γq
√S = 63 GeV
(σ_s − σ_HO)/(σ_B + σ_HO)
10^-1
10^-2
n = 0.01
n = n_QCD (p_T)
n = 20
4
8
12
16
p_T

Fig. 4

line in Fig. 3) the ratio is significantly smaller than for $n = n_{QCD}(p_T)$; in the second case (long dashed line) it is significantly larger.

It should be added that in $\tilde{f}(v,w)$ the multitude of terms contribute with almost random signs (some are positive, others negative) and the resulting cancellations add to the suppression of $\sigma_{HO} - \sigma_s$.

Very similar results we have obtained for the contribution of the subprocess $qg \to \gamma q$ to the inclusive cross section for $pp \to \gamma + X$ analyzed also in detail (Fig. 4).[(5)] The same holds also for the contribution of $\gamma q \to \gamma q$ to $\gamma p \to \gamma$ (large p_T) + X (deep inclusive QED Compton). Finally, essentially similar results we have for the contribution of $\gamma\gamma \to q\bar{q}$ to $\gamma\gamma \to$ hadron (large p_T) + X, involving the fragmentation function $q \to$ hadron.

We may then state the following general assertion: **For processes involving structure functions, and/or fragmentation functions, as p_T increases for fixed s, the relative contribution of the part $\tilde{f}(v,w)$ of HOC is suppressed more and more; and the amount of suppression increases with the softness of the structure and/or fragmentation functions.**

3. DETERMINATION OF THE BREMS CONTRIBUTIONS TO THE DOMINANT PART

We proceed to show that the gluon Brems contributions to the dominant part $f_s(v,w)$ can be determined without recourse to the complete calculation. Again, we outline our approach considering the difference (2.1).

Taking as example the Born unitarity graph Fig. 2′(a), the Born contribution $|M_B|^2$ summed over spins and colors is of the form

$$|M_B|^2 \sim \frac{T_B(p_1,p_2)}{[(q-p_2)^2]^2} \tag{3.1}$$

where $T_B(p_1, p_2)$ is a Born trace. Now, considering the interference term of the Brems amplitudes M_1 and M_2 (unitarity graph Fig. 2′(b)), which gives a contribution to HOC, we have

$$M_1 M_2^+ \sim \frac{T_2(p_1, p_2, k)}{(p_1 - k)^2 (q - p_2)^2 (k - p_2)^2 (q - p_2 + k)^2} \tag{3.2}$$

where $T_2(p_1, p_2, k)$ another trace.

We introduce Sudakov variables:

$$k = \alpha p_1 + \beta p_2 + \ell \qquad p_1 \cdot \ell = p_2 \cdot \ell = 0 \qquad \ell = (0; \vec{\ell}_T) \tag{3.3}$$

so that

$$(p_1 - k)^2 = -\beta s \qquad (k - p_2)^2 = -\alpha s$$

Also, we obtain

$$T_2(p_1 \cdot p_2, k) \sim (1 - \alpha - \beta) T_B(p_1, p_2) \tag{3.4}$$

We use dimensional regularization with $n = 4 - 2\epsilon$, $\epsilon < 0$. Then the phase-space of the emitted gluon becomes

$$\int \frac{d^{n-1}k}{2k_0} \delta_+((p_1 + p_2 - q - k)^2) \sim \int_0^1 d\beta \beta^{-\epsilon} \delta(\beta - \frac{v(1-w)}{1 - vw}) \int_0^{1-\beta} d\alpha \alpha^{-\epsilon};$$

notice that for $w \to 1 : \beta \sim 1 - w$. Next, we consider the leading contribution for $\alpha \to 0$ (i.e. $k \simeq \beta p_2$), and we obtain:

$$\frac{d\sigma}{dv dw} \sim \int \frac{d^{n-1}k}{2k_0} \delta_+((p_1 + p_2 - q - k)^2 M_1 M_2^+$$

$$\sim \int d\beta \beta^{-\epsilon} \delta(\beta - \frac{v(1-w)}{1 - vw}) \int d\alpha \alpha^{-\epsilon} \frac{1}{\alpha\beta} \frac{T_B(p_1, (1-\beta)p_2)}{[(q - (1-\beta)p_2)^2]^2}$$

$$\sim \int d\beta \beta^{-1-\epsilon} \delta(\beta - \frac{v(1-w)}{1 - vw}) |M_B(p_1, (1-\beta)p_2)|^2 \int d\alpha \alpha^{-1-\epsilon} \tag{3.5}$$

Notice that $|M_B|^2$ appears with argument $p_1, (1-\beta)p_2$, as it is pertinent to the emission of a gluon with $k \simeq \beta p_2$, thus expressing a factorization property.

Now $\int d\alpha \alpha^{-1-\epsilon} \sim -1/\epsilon$, and for $w \to 1 : 1-\beta = (1-v)/(1-vw) \to 1$ so that

$$\frac{d\sigma}{dvdw} \sim \left(\frac{v(1-w)}{1-vw}\right)^{-1-\epsilon} \cdot \left(-\frac{1}{\epsilon}\right)|M_B(p_1,p_2)|^2$$

and for $w \to 1$:

$$\frac{d\sigma}{dvdw} \sim \left(\frac{v}{1-v}\right)^{-\epsilon}(1-w)^{-1-\epsilon}\left(-\frac{1}{\epsilon}\right)|M_B(p_1,p_2)|^2 \tag{3.6}$$

We use the expansions

$$\left(\frac{v}{1-v}\right)^{-\epsilon} = 1 - \epsilon \ell n \frac{v}{1-v} + 0(\epsilon^2)$$

and

$$(1-w)^{-1-\epsilon} = -\frac{1}{\epsilon}\delta(1-w) + \frac{1}{(1-w)_+} - \epsilon\left(\frac{\ell n(1-w)}{1-w}\right)_+ + 0(\epsilon^2) \tag{3.7}$$

We write $|M_B(p_1,p_2)|^2 \sim B(v) = v^2 + (1-v)^2$, and we obtain:

$$\frac{d\sigma}{dvdw} \sim \Big[\frac{1}{\epsilon^2}\delta(1-w) - \frac{1}{\epsilon}\frac{1}{(1-w)_+} - \frac{1}{\epsilon}\delta(1-w)\ell n\frac{v}{1-v}$$

$$+\frac{1}{(1-w)_+}(\ell n v - \ell n(1-v)) + \left(\frac{\ell n(1-w)}{1-w}\right)_+ + \frac{1}{2}\delta(1-w)\ell n^2\frac{v}{1-v}\Big]B(v) \tag{3.8}$$

The term $(1/\epsilon^2)\delta(1-w)$ represents an infrared singularity and is cancelled by a similar term of a proper loop graph (Bloch–Nordsieck mechanism). The term $(1/\epsilon)\cdot(1/(1-w)_+)$ is proportional to part of the Altarelli–Parisi split function $P_{qq}(w)$; it is absorbed in the redefinition of the parton distributions (factorization procedure). The third term is cancelled by a similar term arising from a loop graph. The last three terms amount to finite pieces and contribute to $b(v)$, $c(v)$ and $a(v)$ (Eq. (2.4b)) respectively;

it can be seen by inspection of the Appendix of Ref. 3 that indeed terms of this type appear in the expression of the HOC.

4. THE K-FACTOR

We have seen that in $f_s(v,w)$ which dominates the HOC, the various terms contribute at $w = 1$ or mainly near $w = 1$. Moreover, $b(v)$, $c(v)$ and part of $a(v)$ are proportional to the Born $B(v)$ (see (3.8) as well as the Appendix of Ref. 3). Then we simplify as follows: We neglect $\tilde{f}(v,w)$ and we approximately write[6]

$$f(v,w) \approx C\pi^2 B(v)\delta(1-w) \tag{4.1}$$

Regarding C, it is independent of w and a smooth function of v (see e.g. (3.8)); as a first approximation we take $C \sim$ const. Then

$$\begin{aligned} \frac{d\sigma}{dv\,dw} &\sim B(v)\delta(1-w) + \frac{\alpha_s(p_T)}{\pi} f(v,w)\theta(1-w) \\ &\sim B(v)\delta(1-w)[1 + \frac{\alpha_s(p_T)}{\pi} C\pi^2] \end{aligned} \tag{4.2}$$

This leads immediately to the K-factor:

$$K \simeq 1 + \frac{\alpha_s(p_T)}{\pi} C\pi^2 \tag{4.3}$$

Clearly this is a slowly varying function of p_T, and this explains the fact that the overall inclusive cross section (Born + HOC) differs from the Born by an almost constant factor. The K-factor (4.3) is precisely of the form studied in Ref. 2.

5. SUPERSYMMETRIC QCD (SQCD)

Regarding SQCD complete HOC have been determined only for the Drell–Yan type subprocesses:[7]

$$q\bar{q} \to \tilde{\gamma}^* \to \ell^+\tilde{\ell}^- \qquad\qquad \tilde{q}\bar{\tilde{q}} \to \gamma^* \to \ell^+\ell^- \tag{5.1}$$

and for the time-reversed

$$\tilde{\gamma}^* \rightarrow q\tilde{\bar{q}} \qquad\qquad e^-e^+ \rightarrow \gamma^* \rightarrow \tilde{q}\tilde{\bar{q}}; \tag{5.2}$$

In fact, Ref. 7 considers ultrahigh energies and very large transfers so that all partons (including $\tilde{q}$) can be treated as massless.[8]

Since we are interested in processes involving structure (or fragmentation) functions we consider the two subprocesses (5.1). We denote by Q the 4-momentum of $\tilde{\gamma}^*$ or γ^* and introduce the usual scaling variable

$$\tau \equiv \frac{Q^2}{s}$$

Then the contribution of $q\tilde{\bar{q}} \rightarrow \ell^+\tilde{\ell}^-$ to the inclusive cross section of the physical process $AB \rightarrow \ell^+\tilde{\ell}^- + X$ (or of $\tilde{q}\tilde{\bar{q}} \rightarrow \ell^+\ell^-$ to $AB \rightarrow \ell^+\ell^- + X$) can be written:

$$\sigma \equiv \frac{d\sigma}{dQ^2} \sim \int \frac{dx_a dx_b}{x_a x_b} F_{a/A}(x_a) F_{b/B}(x_b) \{\delta(1 - \frac{\tau}{x_a x_b})$$
$$+ \frac{\alpha_s(Q^2)}{\pi} f(\frac{\tau}{x_a x_b}) \theta(1 - \frac{\tau}{x_a x_b})\} \tag{5.3}$$

where $\delta(1 - \tau/x_a x_b)$ is the Born term and f stands for the HOC.

The form of f is very similar for the two processes (5.1),[7] and here we present expressions only for the first. In a form similar to (2.4):

$$f(\tau) = f_s(\tau) + \tilde{f}(\tau) \tag{5.4a}$$

where[7]

$$f_s(\tau) = \frac{1}{2} C_F \{\frac{2}{3}\pi^2 \delta(1-\tau) + 8\frac{1}{(1-\tau)_+} + 8(\frac{\ell n(1-\tau)}{1-\tau})_+\} \tag{5.4b}$$

and

$$\tilde{f}(\tau) = C_F \{\frac{3}{2}(1-\tau) - 4 + (1-\tau)\ell n(1-\tau) - 4\ell n(1-\tau)\} \tag{5.4c}$$

Here $\tilde{f}$ is not very complicated, but a, b and c are even simpler – just constants. In fact, $a \sim C_F \pi^2$, the well-known π^2-term due to a gluon loop in the soft gluon limit,[7] which can be extracted without carrying the complete calculation of HOC.[2]

As in Sect. 2, we denote by σ_s the contribution of f_s to the inclusive cross section σ, and by σ_{HO} that of the complete f; thus $\sigma_{HO} - \sigma_s$ is the contribution of $\tilde{f}$. For our SQCD processes we present results for both the ratios

$$\frac{\sigma_{HO} - \sigma_s}{\sigma_{HO}} \qquad \frac{\sigma_{HO} - \sigma_s}{\sigma_{Born} + \sigma_{HO}} \tag{5.5}$$

We consider the physical processes $pp \to \ell^+ \tilde{\ell}^- + X$ and $pp \to \ell^+ \ell^- + X$ at $\sqrt{s} = 40$ TeV. For $F_{a/A}$ and $F_{b/B}$ we use simple forms $\sim (1-x)^n$ and take for the quark $n_q = 3$ and for the squark $n_{\tilde{q}} = 7$; this is sufficient, in particular since we are interested in ratios of cross sections. To determine $\alpha_s(Q^2)$ we use 6 flavors and $\Lambda = 0.2$ GeV.

Fig. 5(a) and (b) presents our results. We see that both ratios (5.5) are small, in particular the second; for $\sqrt{\tau} \gtrsim 0.05$ they decrease with τ.

As for QCD, the explanation lies in the behaviour $\sim (1 - x)^n$ of the structure functions. From (5.3), the contributions to σ_{HO} arise by integrating in the region $x_a x_b \geq \tau$ with $x_a, x_b \leq 1$. The terms $f_s(\tau)$ give their contribution at or mainly near the boundary $x_a x_b = \tau$, and they dominate the HOC.

It should be clear that our general assertion, end of Sect. 2, holds just as well for SQCD subprocesses with massless partons.

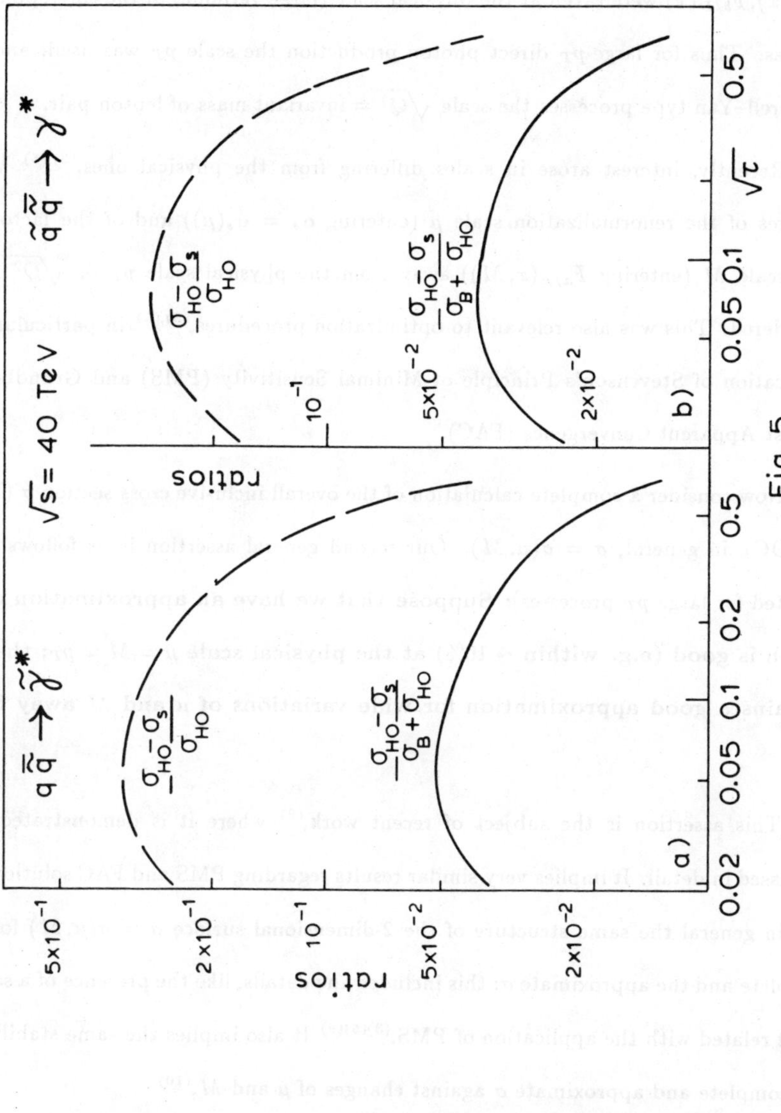

√s = 40 TeV
$q\tilde{q} \to \tilde{\gamma}^*$
$q\tilde{q} \to \tilde{\gamma}^*$
$\frac{\sigma_{HO} - \sigma_s}{\sigma_{HO}}$
$\frac{\sigma_{HO} - \sigma_s}{\sigma_B + \sigma_{HO}}$
ratios
5x10^-1
2x10^-1
10^-1
5x10^-2
2x10^-2
a)
b)
0.02 0.05 0.1 0.2 0.5
0.05 0.1 0.5
$\sqrt{\tau}$

Fig. 5

THE SECOND GENERAL ASSERTION

In all the preceding work the running coupling α_s and the structure functions $F_{a/A}(x), F_{b/B}(x)$ were taken at the physical scale (large variable) of the corresponding process. Thus for large-p_T direct photon production the scale p_T was used; and for the Drell–Yan type processes the scale $\sqrt{Q^2}$ = invariant mass of lepton pair.

Recently, interest arose in scales differing from the physical ones.[(3),(5)] Wide changes of the renormalization scale μ (entering $\alpha_s = \alpha_s(\mu)$) and of the factorization scale M (entering $F_{a/A}(x, M)$) away from the physical scale p_T or $\sqrt{Q^2}$ were considered. This was also relevant to optimization procedures,[(3),(5)] in particular the application of Stevenson's Principle of Minimal Sensitivity (PMS) and Grundberg's Fastest Apparent Convergence (FAC).

Now consider a complete calculation of the overall inclusive cross section σ (Born + HOC); in general, $\sigma \doteq \sigma(\mu, M)$. Our second general assertion is as follows (formulated for large p_T processes): **Suppose that we have an approximation of σ, which is good (e.g. within $\sim 10\%$) at the physical scale $\mu = M = p_T$; then it remains a good approximation for wide variations of μ and M away from p_T.**

This assertion is the subject of recent work,[(9)] where it is demonstrated and discussed in detail. It implies very similar results regarding PMS and FAC solutions,[(9)] and in general the same structure of the 2-dimensional surface $\sigma = \sigma(\mu, M)$ for the complete and the approximate σ; this includes fine details, like the presence of a saddle point related with the application of PMS.[(3)(5)(9)] It also implies the same stability of the complete and approximate σ against changes of μ and M.[(10)]

Clearly, the inclusive cross section σ_s of Sects 2 and 5 is an example of such an approximation; our general assertion holds for σ_s. Another example corresponding to a further (rough) approximation is provided by (4.2) (or (4.3)). Yet, for an approximation of precisely this type, Ref. 9 has demonstrated excellent agreement with the complete $\sigma(\mu, M)$ for wide variations of μ and M.

ACKNOWLEDGEMENTS

We are indebted to S.D. Ellis and A.H. Mueller for criticism of previous work and encouragement and to P. Aurenche et al for making available computer outputs of their complete calculation of HOC for large-p_T direct photon production.

REFERENCES AND FOOTNOTES

1. See e.g. R. Ellis, G. Martinelli, R. Petronzio, Phys. Lett. 104B, 45 (1981) and Nucl. Phys. B211, 106 (1983); R. Ellis, M. Furman, H. Haber, I. Hinchliffe, ibid B173, 397 (1980).
2. A.P. Contogouris, H. Tanaka, Phys. Rev. D33, 1265 (1986).
3. P. Aurenche et al, preprint LPTHE Orsay 87/30.
4. In Figs 3 and 4, $n = n_{QCD}(p_T)$ means parton distributions of Duke–Owens set 1. For the processes of these figures: $\sigma_s > \sigma_{HO}$, so we present the ratio $(\sigma_s - \sigma_{HO})/(\sigma_{Born} + \sigma_{HO})$.
5. We are indebted to P. Aurenche et al for providing computer outputs of their complete calculation of HOC for direct photon production.
6. This form is suggested by the work of Ref. 2 on K-factors.
7. A.P. Contogouris, N. Mebarki, H. Tanaka, Mod. Phys. Lett. A2, 735 (1987).
8. With the exception of $e^- e^+ \to \tilde{q}\bar{\tilde{q}}$ and perhaps of $q\bar{\tilde{q}} \to \ell^+ \tilde{\ell}^-$, the processes (5.1) and (5.2) are not of physical interest. They serve, however, as useful and simple models to study questions of the structure etc of HOC in SQCD.[7]
9. A.P. Contogouris, N. Mebarki, H. Tanaka, Phys. Rev. D37, 2458 (1988).
10. On the basis of Ref. 9, it is easily seen that changes of the renormalization scheme (e.g. $\overline{MS} \to MS$) can be carried for the approximate σ with very similar results as for the complete.

ALTERNATIVE DERIVATION OF THE TWO-LOOP β-FUNCTION OF THE NON-LINEAR SIGMA MODEL IN $2+\epsilon$ DIMENSIONS

M. Leblanc and R.B. Mann
Department of Physics
University of Waterloo
Waterloo, Ontario
CANADA
N2L 3G1

Abstract

Evaluation of the two-loop β-function of the two dimensional non-linear sigma model is carried out in $2+\epsilon$ dimensions using a recently developed scheme called Operator Regularization . The results are shown to be in agreement with those of dimensional regularization.

Introduction

Two-dimensional non-linear sigma-models have a variety of interesting features which have caused them to be a subject of interest to theorists in recent years. These include not only their similiarities to four-dimensional gauge theories[(1)], but also the relationship between their quantum properties and the geometry of the target-space manifold[(2–5)], as well as their connections with string theory[(6–10)]. For example, by requiring the β-function of the sigma-model to vanish, the classical equations of motion of the corresponding string theory are recovered[(6–10)], allowing one to compute tree-level scattering amplitudes for the associated massless fields[(11)].

The non-triviality of the relationship between the properties of sigma-models and the calculational methods used in deriving such properties has also been the subject of recent study. Arguments that the $N = 2$ supersymmetric sigma-model is finite to all orders when the target manifold is Ricci-flat[(12)] have been contradicted by explicit calculations (in the context of dimensional regularization) to four-loop order[(13)]. For the non-linear sigma-model with a Wess-Zumino-Witten (WZW) term there are subtleties involved in treating the ϵ-tensor[(15–18)] in evaluating the β-function to two-loop order when dimensional continuation procedures (such as dimensional regularization) are used. The resultant dependence of the β-function on the choice of prescription for the ϵ-tensor to a certain extent corresponds to field-redefinition ambiguities[(19)]; however there still remain discrepancies between various choices when covariance with respect to the torsion-dependent connection is enforced[(20)].

Regularization of two-dimensional sigma models is inherently problematical because of the two dimensionality of the theory: the conformal group has an infinite number of generators[(21)] and the relationship between infrared and ultraviolet divergences can become blurred. As well, supersymmetry crucially depends upon the number of dimensions. Dimensional continuation methods[(2–5,14–16,22)] therefore pose difficulties because of their continuation away from these special cases.

A new regularization technique called Operator Regularization has recently been introduced by McKeon and Sherry[(23)] . This technique is a means of computing radiative corrections in a given quantum field theory to any given loop order*; at one-loop order it is similar to the zeta-function technique[(24)]. In operator

* By 'loop order'we mean to first order in $\hbar$; see ref. [23] for a discussion on this point in the context of operator regularization.

regularization , the background field quantization formalism is used in conjuction with the Feynman path integral. The path integral is then evaluated prior to an expansion in powers of the background field, in contradistinction to the usual diagrammatic approach. The resultant generating functional is then expressed in terms of determinants and inverses of operators. It is these functions of operators occuring in the expression for the generating functional that are regulated (in a manner described in the next section) as opposed to regulating the theory by the insertion of a regulating parameter into the initial Lagrangian. Hence this regularization does not *a-priori* violate symmetries present in the original action. Remarkably, no explicit divergences arise in this procedure , even as the regulating parameter s approaches its physical value[23].

It is also possible to apply operator regularization to theories in which the original action *is* modified by some other regulating parameter. The final expressions obtained then will depend on both this parameter and the operator regularization parameter s. For example, by writing the action in an arbitrary number of dimensions, n, it is possible to investigate the n-dependence of the quantity under consideration by letting $n \to N$ at the end of the calculation. In this manner one can compare to the results of dimensional regularization using the methods of operator regularization . In contraxt to this, in operator regularization all quantities are evaluated in the *exact* number, N, of space-time dimensions appropriate to the theory. Thus we have

$$\text{operator regularization :} \qquad \Gamma[\phi] = \frac{1}{2} \lim_{s\to 0} F'(s;N)$$

$$\text{dimensional regularization :} \qquad \Gamma[\phi] = \frac{1}{2} \lim_{n\to N} \lim_{s\to 0} F'(s;n)$$

$$= \frac{1}{2} \lim_{\epsilon\to 0} \lim_{s\to 0} F'(s;N+\epsilon).$$

where $F(s,n)$ is an operator functional whose evaluation is described in the next section. The distinction between these two approaches is important. In the latter case the original action is modified by continuing it outside of N dimensions; in operator regularization there is no modification of the original action[23]. Furthermore, in the former case no divergences occur at any step of the procedure, even as the regulating parameter approaches its limiting value; in the latter case there will in general be divergences as $\epsilon \to 0$. Finally, it is important to note that this approach is not diagrammatic; no judicious insertion of the s-parameter in the usual diagrammatic series allows recovery of the operator regularization results[23,25].

These features of Operator Regularization motivated us[26] to apply it to the evaluation of the (one-loop) trace anomaly in the non-linear sigma-model in both 2 and $2+\epsilon$ dimensions in the presence of an arbitrary two-dimensional metric in the weak-field limit. We found that in operator regularization the trace anomaly was uniquely determined in this limit whereas in dimensional regularization its evaluation was ambiguous. This ambiguity may be resolved by appealing to criteria beyond the removal of divergent quantities, such as the relationship between the integrated trace anomaly and the β-function[5,26]. In operator regularization no divergent quantities occur and the result is unique.

In the present work we carry out an evaluation of the β-function of the $2+\epsilon$ dimensional non-linear sigma model with WZW term to two-loop order. In view of the above comments, our chief interest is in using operator regularization to evaluate this quantity in *exactly* 2 dimensions. However, as almost all previous work on this subject has been in the context of dimensional regularization [2–5,13–20], the evaluation of the β-function in the $2+\epsilon$- dimensional case is an interesting problem in its own right, and is a necessary cross-check on the methodology of operator regularization . We shall investigate evaluation of the two-loop beta function in exactly 2 dimensions in a future publication.

2. Operator Regularization

We briefly review the basic technique of operator regularization and then discuss its application to the non-linear sigma model. For a more detailed review of the method, see ref. [25].

Consider a Lagrangian of the form

$$\mathcal{L}[f_i+h_i] = \hbar^2\frac{1}{2}h_iM_{ij}(f)h_j + \frac{\hbar^3}{3!}a_{ijk}(f)h_ih_jh_k + \frac{\hbar^4}{4!}b_{ijk\ell}(f)h_ih_jh_kh_\ell + \cdots \quad (1)$$

where h_i denotes quantum corrections to the background fields f_i[27]. In four-dimensional renormalizable theories higher powers of h_i beyond order 4 do not occur and b is independent the f_i; in the non-linear sigma model, there will be terms to all orders.

The generating functional associated with the Lagrangian of eq. (1) is

$$\begin{aligned} Z[f_i,J_i] &= \int dh_i \exp\left\{\frac{1}{\hbar}\int dx[\mathcal{L}[f_i+h_i]+J_i\hbar h_i]\right\} \\ &= e^{\frac{1}{\hbar}S[f]}\,\mathrm{sdet}^{-\frac{1}{2}}\left[-M_{ij}(f)/\mu^2\right]\sum_{n=0}^{\infty}\frac{\hbar^{-n}}{n!}\left[\int dx\left(\frac{\mu^{-3}}{3!}a_{ijk}(f)\frac{\delta^3}{\delta J_i\delta J_j\delta J_k}\right.\right. \\ &\quad \left.\left. + \frac{\mu^{-4}}{4!}b_{ijk\ell}(f)\frac{\delta^4}{\delta J_i\delta J_j\delta J_k\delta J_\ell}+\cdots\right)\right]^n \exp\left[\frac{\hbar}{2}\int dx J_i(\frac{-M_{ij}(f)}{\mu^2})^{-1}J_j\right] \end{aligned} \quad (2)$$

where $S[f]$ is the classical background action, $dx \equiv d^n x\sqrt{g}$ is the Euclidean measure of the underlying space with metric $g_{\mu\nu}$, and we have rescaled the quantum fields by a factor of $\sqrt{\hbar}$ to remove the dependence of the (super) determinant on $\hbar$ (we shall subsequently set $\hbar = 1$). The parameter μ^2 arises as a consequence of the arbitrariness in the normalization of the functional integral(23,24,26). The superdeterminant is given by the usual ratio(28) of determinants of linear combinations of the bosonic and fermionic parts of $M[f]$. The matrix $M[f]$ is typically a differential operator which acts on bosonic and fermionic fields.

It is these formal expressions of operators that are regulated directly(23) in operator regularization , instead of the eigenvalues of these expressions(24). Formally, for an operator A,

$$\ell n A = -\lim_{s\to 0}\frac{d^n}{ds^n}\left(\frac{s^{n-1}}{n!}A^{-s}\right) \quad (n = 1, 2, ...) \quad . \tag{3}$$

This equation is the fundamental equation used to regulate operators; expressions for both determinants and inverses of operators may be obtained from it by repeated differentiation.

The one-loop generating functional is given by products of determinants of operators; from eqs. (3) we have

$$s\det A = \exp -\zeta_s'(0) \tag{4}$$

where

$$\zeta_s(s) = \frac{1}{\Gamma(s)}\int_0^\infty dt\ t^{s-1} str e^{-At} \quad . \tag{5}$$

In the absence of fermions, the superdeterminant and supertrace are replaced by the usual determinant and trace(23). Evaluating quantities to more than one-loop order involves inverses of operators. Repeated differentiation of (3) leads to

$$\Gamma(N)A^{-N} = \lim_{s\to 0}\frac{d^n}{ds^n}\left[\frac{s^{n-1}}{n!}\frac{\Gamma(s+N)}{\Gamma(s)}A^{-s-N}\right] \quad . \tag{6}$$

Products of inverses of operators are regulated via

$$A_1^{-1}A_2^{-1}...A_p^{-1} = \lim_{s\to 0}\frac{d^n}{ds^n}\left[\frac{s^n}{n!}A_1^{-s-1}A_2^{-s-1}...A_p^{-s-1}\right] \quad . \tag{7}$$

The operator $M[f]$ is then decomposed into a part which is diagonalizable on some space of states and a part which is (perturbatively) interacting ($M =$

$M_0 + M_I$). Using the Schwinger expansions[29]

$$(s)tre^{-(M_0+M_I)t} = (s)tr[e^{-M_0t} + \frac{(-t)}{1}e^{-M_0t}M_I + \frac{(-t)^2}{2}\int_0^1 due^{-(1-u)M_0t}M_1e^{-uM_0t}M_I + ...] \quad (8a)$$

and

$$e^{-(M_0+M_I)t} = e^{-M_0t} + (-t)\int_0^1 due^{-(1-u)M_0t}M_Ie^{-uM_0t} + (-t)^2\int_0^1 du\ u\int_0^1 dv\ e^{-(1-u)M_0t}M_Ie^{-u(1-v)M_0t}M_Ie^{-uvM_0t} + \quad (8b)$$

specific Green's functions may be evaluated as (power-series) functionals of the background field of f_i. The relevant functional traces and matrix elements are generally computed in momentum space using the standard equations[29]

$$< p \mid x > = e^{-ip\cdot x}/(2\pi)^{n/2} \quad (9a)$$

$$< p \mid q > =\equiv \int d^nx\sqrt{g}\frac{e^{-i(p-q)\cdot x}}{(2\pi)^n} \quad (9b)$$

$$< p \mid f(x) \mid q > = \int d^nx\sqrt{g}f(x)\frac{e^{-i(p-q)\cdot x}}{(2\pi)^n} = (\sqrt{g}f)(p-q)/(2\pi)^{n/2} \quad (9c)$$

which result in straightforwardly computable (and non-divergent) integrals. For an operator A,

$$\mathrm{Tr\,A} = \int \mathrm{d^nx}\ \sqrt{\gamma}\langle \mathrm{x|A|x}\rangle \quad (10a)$$

or, in terms of a momentum space states

$$\mathrm{Tr\,A} = \int d^nx\,(\sqrt{g})^{-1}\int d^npd^nq\ \frac{e^{-i(q-p)\cdot x}}{(2\pi)^n}\langle p|A|q\rangle. \quad (10b)$$

3. The Non-Linear Sigma Model

We shall now briefly describe the essential properties of the bosonic non-linear sigma model which are relevant to our calculation.

Consider a non-linear σ-model on a general Riemannian target manifold $\mathcal{M}$ with metric tensor $g_{ij}(\varphi^k)$ where the $\varphi^k(x)$, $k = 1 \dots d$, which are a set of coordinates on $\mathcal{M}$, are taken as fields over a two-dimensional base space with Euclidean signature. The action is given by

$$S[\varphi] = -\frac{1}{2}\int d^2x\ \sqrt{\gamma}\left\{\gamma^{\mu\nu}\partial_\mu\varphi^i\partial_\nu\varphi^j g_{ij}(\varphi) + \frac{2}{3}\eta\epsilon^{\mu\nu}b_{ij}(\varphi)\partial_\mu\varphi^i\partial_\nu\varphi^j\right\} \quad (11)$$

where $\gamma^{\mu\nu}$ is the base space metric and $\gamma = \det\gamma^{\mu\nu}$, $b_{ij} = -b_{ji}$ is the torsion potential on the manifold and $\epsilon_{\mu\nu} = \sqrt{\gamma}\,[\mu,\nu]$ with $[\mu,\nu]$ being the usual (flat space) permutation symbol. In 2 dimensions this action is conformally invariant; in $2+\epsilon$ dimensions it is not conformally invariant in the limit $\epsilon \to 0$.

We expand φ^i in terms of quantum fluctuations, π^i, about a background classical field ϕ^i and, following ref.[2] define an effective action, Γ, by

$$e^{\Gamma} = Z[\phi] = \int [d\pi^i] e^{\{S[\phi+\pi]-S[\phi]-\int d^2x\,\sqrt{\gamma}\pi^i \frac{\delta S}{\delta\phi^i}\}} \tag{12}$$

In order to that our formalism may remain covariant throughout, we shall work in the normal co-ordinate expansion[(2)] where π^i is expressed in terms of a new field, $\xi^i(x)$ which transforms as a vector on $\mathcal{M}$ and will be taken to be the quantum field. ξ^i is defined[(2)] by assuming that ϕ^i and $\phi^i + \pi^i$, are close enough together on the target manifold that they can be joined by a unique geodesic parameterized by $\lambda^i(t)$ where

$$\ddot{\lambda}^i(t) + \Gamma^i_{jk}\dot{\lambda}^j(t)\dot{\lambda}^k(t) = 0 \qquad \cdot \equiv d/dt \tag{13}$$

where t is proportional to arc length and where $\lambda^i(0) = \phi^i$ and $\lambda^i(1) = \phi^i + \pi^i$; ξ^i is then taken to be the tangent vector to the geodesic at $t = 0$, $\xi^i = \dot{\lambda}^i(0)$. By repeatedly differentiating the geodesic equation (13) and applying the boundary conditions we have

$$\lambda^i(t) = \sum_{n=0}^{\infty} \frac{1}{n!}\frac{d^n}{dt^n}\lambda^i(t)\bigg|_{t=0} t^n \tag{14}$$

where

$$\frac{d^n}{dt^n}\lambda^i(0) = -\Gamma^i_{j_1 j_2 \cdots j_n} \equiv -\nabla_{j_1}\Gamma^i_{j_2 j_3 \cdots j_n} = -\nabla_{j_1}\cdots\nabla_{j_{n-2}}\Gamma^i_{j_{n-1}j_n} \tag{15a}$$

and where ∇_j is the covariant derivative with respect to lower indices only. Thus we have

$$\lambda^i(t) = \phi^i + \xi^i t - \frac{1}{2!}\Gamma^i_{j_1 j_2}\xi^{j_1}\xi^{j_2}t^2 - \frac{1}{3!}\Gamma^i_{j_1 j_2 j_3}\xi^{j_1}\xi^{j_2}\xi^{j_3}t^3 \ldots \tag{15b}$$

The normal co-ordinate expansion is defined as follows. It is possible to choose coordinates at the point ϕ so that all coefficients in the above series vanish except for the first two, *ie.*, $\bar{\Gamma}^i_{(j_1\cdots j_n)} = 0$, where an over-bar indicates that a quantity has been evaluated in normal coordinates. In this coordinate system, the geodesic λ^i is a "straight line". This information may be used to obtain a covariant expansion for any tensor; for example a second rank tensor T_{ij} has the expansion

$$T_{kl}(\phi+\pi) = T_{kl}(\phi) + D_i T_{kl}(\phi)\xi^i + \Big(D_{(i_1} D_{i_2)} T_{kl}(\phi)$$

$$+\frac{1}{6} R^n{}_{(i_1 i_2)k}(\phi) T_{nl}(\phi) + \frac{1}{6} R^n{}_{(i_1 i_2)k}(\phi) T_{nl}(\phi) \Big) \xi^{i_1}\xi^{i_2} + O(\xi^3) \tag{16}$$

where the "bars" have been removed from this equation since the right-hand-side is generally covariant and hence valid in any coordinate system.

To evaluate the two-loop β-function we need the action expanded to fourth order in the quantum fields[(2)]. For simplicity we shall choose our two dimensional co-ordinates so that the two-dimensional metric $g_{\mu\nu} = \eta_{\mu\nu}$. Hence to this order the action is[(2,14)]

$$S(\phi_{cl} + \pi(\xi)) = S(\phi_{cl}) + \int d^2x \xi^a [\frac{1}{2}\partial_\mu \partial^\mu \delta_{ab} + N^\mu_{ab}\partial_\mu + \frac{1}{2}\mathcal{M}_{ab}]\xi^b$$

$$+ \int d^2x [\frac{1}{3!} A_{abc}\xi^a\xi^b\xi^c + \frac{1}{2!} B^{abc}_\mu \xi_a\xi_b \mathcal{D}^\mu \xi_c + \frac{1}{2!} C^{abc}_{\mu\nu}\xi_a \mathcal{D}^\mu \xi_b \mathcal{D}^\nu \xi_c] \tag{17}$$

$$+ \int d^2x [\frac{1}{4!} H^{abcd}\xi_a\xi_b\xi_c\xi_d + \frac{1}{4} L^{abcd}_{\mu\nu}\xi_a\xi_b \mathcal{D}^\mu \xi_c \mathcal{D}^\nu \xi_d]$$

where the following quantities are defined:

$$-M^{ab} = p^2\delta^{ab} - iN^{ab}_\mu p^\mu - ip^\mu N^{ab}_\mu - \mathcal{M}^{ab} \tag{18a}$$

with $p^\mu = i\partial^\mu$, and

$$N^{ab}_\mu = \partial_\mu {\phi_{cl}}^k [e^a_i \Gamma^i_{jk} e^{jb} - (\partial_k e^a_i) e^{ib}] - \epsilon_{\mu\nu} S^i_{jk} e^a_i e^{jb} \partial^\nu {\phi_{cl}}^k \tag{18b}$$

$$-\mathcal{M}_{ab} = N_{\mu ca} N^{\mu c}{}_b + \mathcal{R}_{iabj}\partial_\mu \phi^i_{cl}\partial_\nu \phi^j_{cl}(\delta^{\mu\nu} - \epsilon^{\mu\nu}) \tag{18c}$$

$$A^{abc} = -[\mathcal{D}^a \mathcal{R}_j{}^{bcm} + 2\mathcal{R}^{abc}{}_n S^n_{jm}][\delta^{\mu\nu} - \epsilon^{\mu\nu}][\partial_\mu \phi^j_{cl}\partial_\nu \phi^m_{cl}] \tag{18d}$$

$$B^\mu_{abc} = \frac{-2}{3}[\mathcal{R}_{iabc} + \mathcal{R}_{cabi}]\partial^\mu \phi^i_{cl} + \frac{2}{3}[\mathcal{R}_{iabc} - \mathcal{R}_{cabi}]\epsilon^{\nu\mu}\partial_\nu \phi^i_{cl} \tag{18e}$$

$$C^{\mu\nu}_{abc} = -\frac{2}{3} S_{abc}\epsilon^{\mu\nu} \tag{18f}$$

$$\check{C}^{abc} = -\frac{2}{3} S_{abc} \tag{18g}$$

$$H_{abcd} = -[\mathcal{D}_a\mathcal{D}_b\mathcal{R}_{zcdf} + 3\mathcal{R}_{zabg}\mathcal{R}^g{}_{cdf} + \mathcal{R}_{zabg}\mathcal{R}^g{}_{fcd}$$

$$+ 4\mathcal{D}_a\mathcal{R}_{bzcg}S^g{}_{df} + 4\mathcal{R}_{zabg}S^g{}_{cb}S^b{}_{df}][\delta^{\mu\nu} - \epsilon^{\mu\nu}][\partial_\mu \phi^z_{cl}\partial_\nu \phi^f_{cl}] \tag{18h}$$

$$\hat{L}_{abcd} = -\frac{2}{3}R_{cabd} \tag{18i}$$

$$\check{L}_{abcd} = \mathcal{R}_{cabd} \tag{18j}$$

Here $\hat{L}_{abcd}$ and $\check{L}_{abcd}$ are the symmetric part and antisymmetric parts of $L^{\mu\nu}_{abcd}$ respectively.

In (17) the background field, ϕ^i has been chosen to satisfy the classical equations of motion so that terms proportional to $\delta S/\delta\phi^i$ vanish. We have also made use of vielbeines $e_i{}^a$ on the tangent plane(2). These are defined as follows:

$$e^a{}_i e^b{}_j g^{ij} = \delta^{ab} \qquad \Longrightarrow \qquad e^a{}_i e^b{}_j \delta_{ab} = g_{ij} \tag{19}$$

and so $\xi^a \equiv e^a{}_i \xi^i$ (which has the inverse relation $\xi^i = e_a{}^i \xi^a$). $\mathcal{D}_\mu$ is the full covariant derivative

$$\mathcal{D}_\mu \xi^i = D_\mu \xi^i - S^i{}_{jk} \epsilon_{\mu\nu} \partial^\nu \phi^k \xi^j, \tag{20}$$

and

$$\mathcal{R}_{ijkl} = R_{ijkl} + D_k S_{ijl} - D_l S_{ijk} + S_{mik} S^m{}_{lj} - S_{mil} S^m{}_{kj} \tag{21}$$

is the full curvature including the torsion $S_{ijk} = \eta \partial_{[i} e_{jk]}$; R_{ijkl} is the usual Riemann curvature tensor.

3. The Two-loop β-function in $2+\epsilon$ Dimensions

In this section we shall briefly outline how the two-loop β-function may be obtained using operator regularization in $2+\epsilon$ dimensions.

The β-function (to any loop-order) is determined from the coefficients of the simple poles in ϵ. The coefficients of the higher poles obey a set of relations that arise as a consequence of the (unrenormalized) target-space metric being independent of the renormalization scale μ. In general for some target space field Ψ in $2+\epsilon$ dimensions we have(2,30)

$$\Psi^B = \mu^\epsilon \left[\Psi + \sum_{\nu=0}^{\infty} \epsilon^{-\nu} T^{(\nu)}(\Psi) \right] \tag{22}$$

where Ψ^B is the unrenormalized field(s). Requiring Ψ^B to be independent of μ gives(2,30)

$$\beta^\Psi = \epsilon\Psi + (1 + \lambda\frac{\partial}{\partial\lambda}) T^{(1)}(\lambda^{-1}\Psi)\big|_{\lambda=0} \tag{23}$$

which defines the β-function, and

$$(1 + \lambda\frac{\partial}{\partial\lambda}) T^{(\nu+1)}(\lambda^{-1}\Psi)\bigg|_{\lambda=0} =$$

$$\lim_{\eta\to 0}\eta^{-1}\Big[T^{(\nu)}\Big(\Psi+\eta(1+\lambda\frac{\partial}{\partial\lambda})T^{(\nu)}(\lambda^{-1}\Psi)\Big)-T^{(\nu)}\Big(\Psi\Big)\Big]\Big|_{\lambda=0} \tag{24}$$

In general Ψ represents the target space metric g_{ij} and torsion potential b_{ij}; it may be thought of as a vector $\Psi = (g_{ij}, b_{ij}, V)$ where $V(\phi)$ is a scalar potential field. This term does not appear in the original action (11), but must be added in $2+\epsilon$ dimensions to distinguish infrared from ultraviolet divergences. This is usually done[30] by writing $V(\phi) = m^2P(\phi)$. Such a term destroys the conformal invariance of the original action (11), although this invariance is recovered (classically) as $m \to 0$.

Quantum-mechanical recovery of conformal invariance is not so trivial. Invariance under local scale transformations $g_{\mu\nu} \to e^{\Lambda(x)}g_{\mu\nu}$ occurs if the (renormalized) trace anomaly $\frac{2}{\sqrt{\gamma}}\gamma^{\mu\nu}\frac{\delta\Gamma^R}{\delta\gamma^{\mu\nu}}$ vanishes. If Λ is a constant, then a weaker condition holds: only the integrated trace anomaly must vanish. This quantity may be shown to be related to the β-function[2,30]:

$$\int d^nx\sqrt{g}\langle T\rangle = \beta^{\Psi}\frac{\delta\Gamma^R}{\delta\Psi}+V\frac{\delta\Gamma^R}{\delta V} \tag{25}$$

Hence rigid scale invariance (when Λ is constant) holds when the β-function vanishes. There is however, an ambiguity in defining the β-function due to (a) the possibility of adding finite local counterterms to the effective action and (b) diffeomorphisms of the target space manifold. Taking account of these allows one to relate the vanishing of the trace anomaly (*ie., local* scale invariance, or conformal invariance) to a modified version of the β-function[5,30].

In any event, it is necessary to evaluate the β functions. In operator regularization this involves (to two-loop order), finding the coefficient of $\hbar^2$ in (2), using (6-8) to regulate the operator M. However there is a further complication coming from the fact that the background/quantum split is non-linear[30−32]. A naive renormalization of the one-loop effective action will spoil target-space covariance in higher loops: the quantum field ξ is not multiplicatively renormalized. This may be accounted for by carrying out a "double background" field expansion[30]

$$\exp\frac{\Gamma}{\hbar} = \int [d\xi^a]\exp\hbar^{-1}[S(\phi_{cl}+\pi(\hat{\xi}+\xi_{cl}))+\hat{\xi}_aJ^a] \tag{26}$$

where now $\hat{\xi}$ is the quantum field which is integrated over in the functional integral; its renormalization is determined by removing divergences from $\Gamma[\phi,\xi_{cl}]$.

From the decomposition of $S(\phi_{cl}+\pi(\hat{\xi}+\xi_{cl}))$, we see that $\mathcal{L}, \mathcal{N}^\mu, \mathcal{M}^{ab}$, and the ξ propagator get renormalized after one loop[15,30]. This renormalization produces a new action that can be put in the formulation (26) once again and gives two loop contributions to the effective action (coming from one loop counterterms). From the determinant piece we get, in $2+\epsilon$ dimensions,

$$\Gamma_{\text{counterterms}} = \frac{\hbar^2}{4\pi\epsilon}[I_1 - m^2 I_2]\int d^n x[\frac{1}{2}\hat{L}_{ccab} - \frac{9}{4}\check{C}_{acd}\check{C}_b^{cd}]\mathcal{M}^{ab}$$
$$-\frac{\hbar^2}{4\pi\epsilon}[I_1 - m^2 I_2]\int d^n x\frac{\epsilon^{\mu\nu}}{n}\check{L}_{ccab}\partial_\mu N_\nu{}^{ab} \tag{27}$$
$$-\frac{\hbar^2}{32\pi^2\epsilon}\int d^n x \mathcal{M}_{ab}\hat{L}^{abc}{}_c + \text{more double pole}$$

This last line in eq.(27) was missing in eq.(28) of ref.[15]. We will see later that the contribution going like $m^2 I_2$ will remove all the "infrared pieces" in the pure 2 loop contributions.

The regulated version of $-M^{-1}$ is:

$$< x| - M_{ab}^{-s-1}|y> = \int \frac{d^n p}{(2\pi)^n}\frac{\delta^{ab}e^{ip(x-y)}}{(p^2+m^2)^{1+s}}$$
$$+\int \frac{d^n p d^n q}{(2\pi)^{3n/2}}e^{i(px-qy)}(s+1)\int_0^1 du\frac{[i(p+q)N_{ab}(p-q)+\mathcal{M}_{ab}(p-q)]}{[(1-u)p^2+uq^2+m^2]^{2+s}} \tag{28}$$

We then employ this formula in (2) to find six contributions to the divergent part of the effective action to order $\hbar^2$. A simplifying feature of the calculation is that it is possible to set the external momenta to zero for any terms that do not depend on the external momenta[14]. Terms which do depend on external momenta (such at Γ_{CC}) may be evaluated by looking at the lowest order terms in the Taylor series in powers of the external momenta as we are only interested in the divergent part of the effective action. We find:

$$\Gamma_H = \frac{d^2}{ds^2}\frac{s^2}{2!}\frac{1}{8}\int d^n x H_{aabb}\Big[\int\frac{d^n p}{(2\pi)^n}\frac{1}{(p^2+m^2)^{1+s}}\Big]^2 \tag{29}$$

$$\Gamma_{ML} = \frac{d^2}{ds^2}\frac{s^2}{2!}\frac{1}{4}\Big[\int\frac{d^n p}{(2\pi)^n}\frac{1}{(p^2+m^2)^{1+s}}\Big]$$
$$\times\Big[\int\frac{d^n r}{(2\pi)^n}\frac{(1+s)r^2}{(r^2+m^2)^{2+s}}\Big]\int d^n q\hat{L}_{ccab}(q)\mathcal{M}^{ab}(-q)$$
$$+\frac{d^2}{ds^2}\frac{s^2}{2!}\frac{1}{4}\Big[\int\frac{d^n p}{(2\pi)^n}\frac{p^2}{(p^2+m^2)^{1+s}}\Big]$$
$$\times\Big[\int\frac{d^n r}{(2\pi)^n}\frac{(1+s)}{(r^2+m^2)^{2+s}}\Big]\int d^n q\hat{L}_{abcc}(q)\mathcal{M}^{ab}(-q) \tag{30}$$

$$\Gamma_{NL} = \frac{d^2}{ds^2}\frac{s^2}{2!}\frac{1}{2n}\left[\int \frac{d^n p}{(2\pi)^n}\frac{1}{(p^2+m^2)^{1+s}}\right]\left[\int \frac{d^n r}{(2\pi)^n}\frac{(1+s)r^2}{(r^2+m^2)^{2+s}}\right]$$
$$\times \int d^n q (iq_\mu) N_\nu{}^{ab}(q)\epsilon^{\mu\nu}\check{L}_{ccab}(-q) \tag{31}$$

$$\Gamma_{BB} = \frac{d^2}{ds^2}\frac{s^2}{2!}\frac{1}{4}\int \frac{d^n p d^n q}{(2\pi)^{2n}}\frac{p_\mu p^\nu - p_\mu q^\nu}{(p^2+m^2)^{1+s}(q^2+m^2)^{1+s}((p+q)^2+m^2)^{1+s}}$$
$$\times \int d^n r B^\mu_{abc} B^{abc}_\nu \tag{32}$$

$$\Gamma_{CC} = \frac{d^2}{ds^2}\frac{s^2}{2!}\frac{(-1-s)}{4n}\int \frac{d^n p d^n q}{(2\pi)^{2n}}\Big[\frac{p_\nu p_\sigma q_\lambda q_\mu(-8+3n-6s)}{(p^2+m^2)^{1+s}(q^2+m^2)^{1+s}((p+q)^2+m^2)^{2+s}}$$
$$+ \frac{p_\nu p_\sigma q_\lambda q_\mu (12+6s)m^2}{(p^2+m^2)^{1+s}(q^2+m^2)^{1+s}((p+q)^2+m^2)^{3+s}}\Big]\int d^n r r^2 \check{C}_{abc}\check{C}^{abc}\epsilon^{\mu\nu}\epsilon^{\lambda\sigma} \tag{33}$$

$$\Gamma_{CCM} = \frac{d^2}{ds^2}\frac{s^2}{2!}\frac{9(1+s)}{4}\int \frac{d^n p d^n q}{(2\pi)^{2n}}\frac{p_\nu p_\lambda q_\mu q_\sigma}{(p^2+m^2)^{1+s}(q^2+m^2)^{1+s}((p+q)^2+m^2)^{2+s}}$$
$$\times \int d^n r \check{C}_{cda}\check{C}^{cd}{}_b \mathcal{M}^{ab}\epsilon^{\nu\mu}\epsilon^{\lambda\sigma} \tag{34}$$

Taking the limit $s \to 0$ followed by the limit $\epsilon \to 0$, where $n = 2+\epsilon$ gives

$$\Gamma_H = \frac{1}{8}I_1^2 H_{aabb} \tag{35}$$

$$\Gamma_{ML} = \frac{1}{4}[(I_1^2 - m^2 I_1 I_2)\hat{L}_{ccab}\mathcal{M}^{ab} - (m^2 I_1 I_2)\hat{L}_{abcc}\mathcal{M}^{ab}] \tag{36}$$

$$\Gamma_{NL} = \frac{1}{4}I_1^2(\mathcal{D}_\mu N_\nu)_{ab}\epsilon^{\mu\nu}\check{L}_{ccab} \tag{37}$$

$$\Gamma_{BB} = \frac{3}{8n}I_1^2 B^\mu_{abc}B^{abc}_\mu \tag{38}$$

$$\Gamma_{CC} = \frac{-1}{16}I_1^2\frac{(5n-8)}{n(n-1)}g(n)\mathcal{D}_\mu\check{C}_{abc}\mathcal{D}^\mu\check{C}^{abc} \tag{39}$$

$$\Gamma_{CCM} = \frac{-9}{16}\frac{g(n)}{(n-1)}(I_1^2 - 2m^2 I_1 I_2)\check{C}_{acd}\check{C}_b{}^{cd}\mathcal{M}^{ab} \tag{40}$$

Which agrees with the results of ref.[15]. In the above formulas an integration $\int d^2x$ is understood. We have also used the prescription

$$\delta_{\rho\sigma}\epsilon^{\mu\rho}\epsilon^{\nu\sigma} = -g(n)\delta^{\mu\nu}; \qquad g(n) = 1 + c\epsilon + O(\epsilon^2) \tag{41}$$

for continuing the product of 2 ϵ-tensors. Also,

$$I_1 = \frac{-1}{2\pi\epsilon} + finite; \qquad m^2 I_2 = \frac{1}{4\pi} \tag{42}$$

If instead one uses prescription[33]

$$\epsilon^{\mu\nu}\epsilon^{\lambda\sigma} = (\delta^{\mu\sigma}\delta^{\nu\lambda} - \delta^{\mu\lambda}\delta^{\nu\sigma})(1+\lambda\epsilon) \tag{43}$$

then

$$\Gamma_{CCM} = \frac{-9}{16}(I_1^2 - 2m^2 I_1 I_2)(1+\lambda\epsilon) \tag{44}$$

which agrees with the above result for λ= c-1. Also,

$$\Gamma_{CC} = \frac{-1}{16}(I_1^2 - 4m^2 I_1 I_2)(1+\lambda\epsilon) \tag{45}$$

which also agrees with the results of ref. [15] for $\lambda = c-1$.

4. Conclusions

We have evaluated the divergent part of the effective action of the $2+\epsilon$-dimensional bosonic non-linear sigma model with torsion to two-loop order using operator regularization . Our results agree with those of Hull and Townsend[15], and Allen and Jones[20] (for c=0). We find all the divergent coefficients to agree with their results. We also find that the prescription (41) for the ϵ-tensor employed in ref. [15] trivially agrees with that of ref. [33] (eq. (43)) in $2+\epsilon$ dimensions provided $\lambda = c-1$. (Allen and Jones worked in $2-\epsilon$ dimensions; in this case one obtains $\lambda = c+1$). Employing the definitions of (18) and summing up all the $1/\epsilon$ poles in (27),(35-40) gives the β-functions (order $\hbar^2$ only, and η =3/2):

$$\beta_{(ij)}(g^R) = -(\frac{\hbar}{2\pi})^2(\frac{3+c}{6})[\mathcal{R}_{iklm}(\mathcal{R}^{klm}{}_j - \frac{1}{2}(\frac{3-c}{3+c})\mathcal{R}^{lmk}{}_j) + \frac{6(1-c)}{3+c}S_{abc}S_{dbc}\mathcal{R}_{iadj}] \tag{46}$$

$$\beta_{[ij]}(b^R) = (\frac{\hbar}{2\pi})^2(\frac{3+c}{6})[\mathcal{R}_{iklm}(\mathcal{R}^{klm}{}_j - \frac{1}{2}(\frac{3-c}{3+c})\mathcal{R}^{lmk}{}_j) + \frac{6(1-c)}{3+c}S_{abc}S_{dbc}\mathcal{R}_{iadj}] \tag{47}$$

The c-dependence (or, alternatively the λ-dependence) can be understood as a field redefinition ambiguity[19,20,33]

Of course it is of much greater interest to employ operator regularization to evaluate these quantities in *exactly* two dimensions. This is particularly interesting for the β-function in the bosonic model. In the $n = 2 + \epsilon$ dimensional case, the curvature squared contribution in (46) arises from $1/n$ times the pure double pole in ϵ in (38). The factor of $1/n$ in turn arises specifically as a consequence of carrying out the calculation in n-dimensions. This factor is exactly 2 in the two dimensional model. The implications for the β function will be explored in a future publication.

Acknowledgements

This work was supported by the Natural Sciences and Engineering Research Council of Canada and by le Fonds F.C.A.R du Québec.

References

1. A.M. Polyakov, Phys. Lett. **B59** (1975) 76; E. Brezin and J. Zinn-Justin, Phys. Rev. Lett.**36** (1976) 691; W. Bardeen, B. Lee, and R. Schrock , Phys. Rev.D14 (1976) 985.
2. L. Alvarez-Gaume, D.Z.Freedman, S.Mukhi. Ann. Phys. **134** (1981) 85-109.
3. D. Freidan, Phys. Rev. Lett.**45** (1980) 1057.
4. E.Braaten, T.L. Curtright, C.K. Zachos. Nucl. Phys. B260, (1985) 630-688.
5. C.M.Hull and P.K.Townsend, Nucl.Phys B274 (1986) 349-362.
6. C. Lovelace, Phys. Lett.**B135** (1984) 75; Nucl. Phys.**B273** (1986) 413.
7. C.G. Callan, D. Freidan, E.J. Martinec and M.J. Perry, Nucl. Phys.**B262** (1985) 593.
8. A. Sen, Phys. Rev.**D32** (1985) 2102; Phys. Rev. Lett.**55** (1985) 1846.
9. B.E. Fridling and A. Jevicki, Phys. Lett.**B174** (1986) 75.
10. C.M. Hull and P. Townsend, Phys. Lett.**B167** (1986) 51.
11. C.G. Callan, I. Klebanov and M.J. Perry, Nucl. Phys.**B278** (1986) 78.
12. L. Alvarez-Gaume, S. Coleman and P. Ginsparg, Comm. Math. Phys. **102** (1985) 311.
13. M.T. Grisaru, A.E.M. van de Ven, and D. Zanon, Phys. Lett.**B173** (1986) 423; Nucl. Phys.**B277** (1986) 388; 409.
14. B.E.Fridling and A.E.M Van De Ven, Nucl.Phys. B268 (1986) 719-736 .
15. C.M.Hull and P.K.Townsend, Phys. Lett.**B191** (1987) 115.
16. E. Guadagnini and M. Mintchev, IPUP-TH 21/86 Pisa Preprint (1986).
17. M. Bos, Phys. Lett.**B189** (1987) 435.

18. D. Zanon, Phys. Lett.**B191** (1987) 363.
19. D.R.T. Jones, Phys. Lett.**B192** (1987) 391; I. Jack and D.R.T. Jones, Phys. Lett.**B193** (1987) 449.
20. R.W. Allen and D.R.T. Jones, Nucl. Phys.**B** (to be published).
21. See for example S. Shenker in *Unified String Theories*, eds. M. Green and D. Gross, World Scientific (1986).
22. D.M. Capper, D.R.T. Jones and P. van Nieuwenhuizen, Nucl. Phys.**B167** (1980) 479.
23. D.G.C Mckeon, T.N. Sherry, Phys. Rev. Lett.**59** (1987) 532; Phys. Rev.**D35** (1987) 3584; Can. J. Phys. (in press).
24. J. Dowker and R. Critchley, Phys. Rev.**D13** (1976) 3224; S. Hawking, Comm. Math. Phys. **55** (1977) 133; A. Salam and J. Strathdee, Nucl. Phys.**90** (1975) 203.
25. R.B. Mann, *Proceedings of the NSERC-CAP Summer Institute in Theoretical Physics*, eds. F.C. Khanna, G. Kunstatter, H.C. Lee and H. Umezawa, Edmonton, Alberta, World Scientific (1987).
26. M. Leblanc, R.B. Mann and B. Shadwick, Phys. Rev.**D37** (1988) 3548.
27. B. de Witt, Phys. Rev. **162** (1967) 1195; L. Abbott, Nucl. Phys.**B185** (1981) 189.
28. P. van Nieuwenhuizen, Phys. Rep. **68** (1981) 189.
29. J. Schwinger, Phys. Rev.**82** (1951) 664.
30. C. Hull in *Super Field Theories*, eds. H.C. Lee, V. Elias, G. Kunstatter, R.B. Mann and K. Viswanathan (NATO ASI series vol. 160 1987).
31. P. Howe, G. Papadopolous and K. Stelle, Nucl. Phys.**296** (1988) 26.
32. A. Blasi, F. Deldue and S.P. Sorella, CERN preprint TH.5046/88 (May 1988).
33. R.R. Metsayev and A.A. Tseytlin, Phys. Lett.**B191** (1987) **354**; Nucl. Phys.**B293** (1987) 385.

M_Q DEPENDENCE OF THE DECAY CONSTANTS OF PSEUDOSCALAR $Q\bar{q}$ MESONS*

Howard D. Trottier[†‡]

Center for Theoretical Physics
Laboratory for Nuclear Science
and Department of Physics
Massachusetts Institute of Technology
Cambridge, Massachusetts 02139 U.S.A.

Roberto R. Mendel[†]

Department of Applied Mathematics
The University of Western Ontario
London, Ontario, Canada N6A 5B9

ABSTRACT

We find that the asymptotic behavior ($r \to 0$) of the upper component of the Dirac wavefunction for a particle in a QCD leading-log-corrected Coulomb potential is $\chi(r \to 0) = \chi(0)[1 - 1/\{-18\pi^2 b_0^2 \log(r\Lambda_\chi)\} + O\{1/\log^3(r\Lambda_\chi)\}]$, where b_0 is the one-loop β-function coefficient, Λ_χ is a scale mass that is expressed in terms of the QCD scale parameter $\Lambda_{\overline{MS}}$, and where $\chi(0)$ is finite. This differs from the singular behavior found by L. Durand for a relativistic spin-0 equation. We then attempt to make a (nearly) model-independent estimate of the M_Q dependence of the pseudoscalar decay constant $f_P(M_Q)$ for $Q\bar{q}$ mesons in the region $M_Q \gtrsim 0.8$ GeV. We find that relativistic and short-distance QCD effects on f_P are very significant. Our results are consistent with recent QCD sum rule and lattice calculations, and indicate that the maximum of $f_P(M_Q)$ is not far from M_c, and that the ratio f_B/f_D is significantly closer to 1 than the naive expectation $f_B/f_D \approx (M_D/M_B)^{1/2} \approx 0.6$. An additional logarithmic correction recently put forward by Shifman and Voloshin (renormalization effect) would further enhance this effect.

---

* Invited talk presented at the **TENTH ANNUAL MONTREAL-ROCHESTER-SYRACUSE-TORONTO MEETING ON HIGH ENERGY THEORY**, University of Toronto, Toronto, Canada, May 9–10, 1988; and MIT preprint CTP#1614, June 1988.

† This work is supported in part by the Natural Sciences and Engineering Research Council of Canada (NSERC).

‡ NSERC Postdoctoral fellow. This work also supported in part by funds provided by the U.S. Department of Energy under contract #DE-AC02-76ER03069.

I. INTRODUCTION

The weak decay constants f_P of the pseudoscalar $Q\bar{q}$ mesons ($D_{u,d,s}$; $B_{u,d,s}$; $T_{u,d,s}$; etc.) have been estimated in the context of many QCD-motivated approximations.[1–11] The predictions for each of these constants cover a wide range of values, sometimes differing by factors of 3 or 4 from one model to another. The various models can also be classified into groups that predict extremely different behaviors for the evolution of the decay constant with increasing heavy quark mass. For example, many non-relativistic quark model calculations[1–3] suggest that $f_K > f_D > f_B$, while the latest QCD sum rule[5–7] and lattice[8–10] calculations suggest that f_P is roughly constant between the K and B mesons. This is a clear indication that our understanding of the basic physics involved is far from satisfactory. Unfortunately, the experimental data for the decay constants is scarce. Only f_π and f_K are accurately known, while an upper limit for f_D has recently been reported by Mark III.[12]

Phenomenologically it is also important to have reliable estimates of these decay constants, because they appear in many processes from which we can learn about quantities of fundamental importance to the standard model, such as the quark mixing matrix, CP violation, the mass of the top quark, etc. (e.g. $B_u \to l\bar{\nu}$, B_d–$\overline{B}_d$ mixing, B_s–$\overline{B}_s$ mixing, etc.).

In this article, we put forward some fairly model-independent arguments that should lead us to a correct qualitative understanding of the behavior (M_Q dependence) of the decay constants $f_P(M_Q)$ of pseudoscalar $Q\bar{q}$ mesons. We make completely model-independent calculations for the behavior of f_P in the asymptotic region $M_Q \gg \Lambda_{\overline{MS}}$ (where $\Lambda_{\overline{MS}}$ is the QCD scale parameter in the modified-minimal-subtraction scheme), but find that the most interesting effects are to be found in the region corresponding to meson masses M_P in the range $M_K \lesssim M_P \lesssim M_B$, where model-dependent factors are more difficult to eliminate. Despite this, we are able to show that two *model-independent dynamical effects*, namely the short-distance (asymptotic) behavior of the QCD running coupling, and relativistic dynamics for the light quark, are responsible for *significant* deviations from the naive non-relativistic result $f_P \propto 1/\sqrt{M_P}$. These effects apparently have either not been included, or understood, in previous calculations.

The main steps used in our analysis of the asymptotic behavior of f_P are as follows:
i) To calculate the light quark wavefunction, the heavy quark is assumed to be "fixed" at the origin. Its recoil and effects due to its color-magnetic moment are neglected in first approximation.
ii) The light quark is a Dirac fermion with a small *current* mass. In the vicinity of the

heavy quark it experiences a pure leading-logarithm corrected Coulomb potential.
iii) To account for the quantum-mechanical fluctuations in the position of the heavy quark, the overlap for the annihilation of the $Q\bar{q}$ pair is averaged over the Compton wavelength of the heavy quark, $1/M_Q$.

The validity of these approximations, especially steps *i*) and *iii*), is discussed in section V and in an Appendix.

The rest of this paper is organized as follows. In Section II we describe the general framework for our calculation of f_P, which is expressed in terms of the light quark wavefunction $\psi(\mathbf{r})$. In section III we attempt to isolate the influence of short-distance and relativistic effects on $f_P(M_Q)$ in a *model-independent* way, by studying the Dirac equation in a pure leading-log-corrected Coulomb potential. Unfortunately, to calculate the decay constants in the region of greatest interest we are required to evaluate the light quark wavefunction $\psi(\mathbf{r})$ at distances where long-distance properties of the $Q\bar{q}$ system may become important; in particular, as we go below $M_Q \approx 1$ GeV, we get dangerously close to the scale at which the leading-log coupling constant blows up.

In order to check that we can isolate true short-distance and relativistic effects from possible spurious effects due to the failure of the leading-log potential at long distances, we introduce a renormalization-group-improved effective action model (EAM) of QCD in section IV, which should describe the $Q\bar{q}$ system at essentially *all* length scales. This effective action model was first proposed by Matinyan and Savvidy, [13] and Pagels and Tomboulis, [14] and was then extensively developed by Adler. [15] This model has many of the qualitative features expected of *exact* QCD; in particular, this model naturally interpolates between asymptotic freedom and infrared slavery (linear confinement). [15] Furthermore, Adler and Piran [16,17] have shown that this model gives a good quantitative account of the quarkonium ($Q\bar{Q}$) potential at all length scales. In earlier work, [18] we extended Adler's model to $Q\bar{q}$ and Qqq hadrons. We will show in a forthcoming publication [19] that an improved version of the model gives a good quantitative account of the spectrum of all lowest-lying hadrons with one heavy quark. We therefore feel confident in using this model to estimate the pseudoscalar decay constants $f_P(M_Q)$.

In section V we compare the results for $f_P(M_Q)$ obtained in the effective action model with the results obtained using the pure leading-log potential. We find that the two approaches give essentially the same results for f_P. We conclude that the EAM can be used to make a *model-independent* estimate of short-distance and relativistic effects. We analyze these effects in detail in section V, and find that they are *very* significant. Our quantitative predictions are in basic agreement with the latest QCD sum rule [5–7] and lattice [8–10] calculations, and have important consequences for phenomenology.

Finally in section VI we summarize our main results and briefly discuss possible future developments and applications of this work. We also briefly mention how our results would be affected by an additional short-distance effect recently put forward by Shifman and Voloshin.[20]

II. GENERAL FRAMEWORK

The pseudoscalar decay constant f_P is defined by the matrix element of the axial vector current A^5_μ between a pseudoscalar meson state $|P_s(K)\rangle$ of definite 4-momentum K and the vacuum[21]

$$\langle 0 \,|\, A^5_\mu(X) \,|\, P_s(K)\rangle \;=\; i p_\mu f_P e^{-iK\cdot X} \,. \qquad (1)$$

f_P enters into the familiar expression for the leptonic decay modes of the charged pseudoscalar[21]

$$\Gamma(P \to l\bar{\nu}) \;=\; f_P^2 \frac{G_F^2}{8\pi} M_P m_l^2 |V_{Q\bar{q}}|^2 \left[1 - \frac{m_l^2}{M_P{}^2}\right]^2 , \qquad (2)$$

where $V_{Q\bar{q}}$ denotes the appropriate Kobayashi-Maskawa matrix element. Note that the normalization of Eq.(1) is such that $f_\pi = 132$ MeV.

In the quark-model picture of the pseudoscalar mesons that we are using here, f_P is related to the amplitude for annihilation of the $Q\bar{q}$ pair, and can be expressed in terms of the wavefunction of the $Q\bar{q}$ system. As described in the Introduction, we assume to begin with that the heavy quark is "fixed" at the origin. The light quark is described by a Dirac wavefunction $\psi(\mathbf{r})$. In the ground state, we expect $\psi(\mathbf{r})$ to have the familiar form (as appropriate for a central $Q\bar{q}$ potential)

$$\psi \;=\; \mathcal{N} \begin{bmatrix} -i\chi(r) \\ \boldsymbol{\sigma}\cdot\hat{\mathbf{r}}\,\phi(r) \end{bmatrix} , \qquad (3)$$

where $\mathcal{N}$ is a normalization constant.

In the limit where the heavy quark has infinite mass, the annihilation takes place at $r = 0$, and f_P is proportional to the upper component at the origin[22]

$$f_{\rm rel}(M_Q = \infty) \;=\; \sqrt{\frac{12}{M_P}}\, \mathcal{N}\, \chi(r=0) \,, \qquad (4)$$

where $\sqrt{12}$ is a color-spin-flavor coefficient, and where the factor of $1/\sqrt{M_P}$ has a kinematical origin [coming from the density of states when the momentum-space meson state vector in Eq.(1) is fourier-transformed to get f_P in terms of the co-ordinate space wavefunction]. Note that we have emphasized that Eq.(4) applies to a *relativistic* light quark by using the notation $f_{\rm rel}$.

Since it is unphysical to localize a heavy quark of finite mass within distances much smaller than its Compton wavelength, $1/M_Q$, Eq.(4) should not be used when M_Q is finite. To try to account for the quantum-mechanical fluctuations in the position of the heavy quark, we consider it to be described by a non-relativistic wavefunction $\Psi_Q(\mathbf{r})$ which is "spread" over a distance r_Q of $O(1/M_Q)$. Then the decay constant $f_{\rm rel}$ is given by the following expression [23]

$$f^2_{\rm rel}(M_Q) \quad = \quad \frac{12}{M_P}\mathcal{N}^2 \int d^3\mathbf{r}\, |\Psi_Q(\mathbf{r})|^2\, \chi^2(r)\ , \tag{5}$$

which reduces to Eq.(4) when $|\Psi_Q(\mathbf{r})|^2 = \delta^3(\mathbf{r})$, as appropriate for the limit $M_Q \to \infty$.

In the applications which we consider in this paper $f_{\rm rel}$ will depend on the "spread" r_Q in the heavy quark wavefunction, but will turn out to be essentially independent of the exact *shape* of $\Psi_Q(\mathbf{r})$. Then Eq.(5) for $f_{\rm rel}$ will be essentially equivalent to the following (physically reasonable) expression

$$f_{\rm rel}(M_Q) \quad = \quad \sqrt{\frac{12}{M_P}}\,\mathcal{N}\,\chi(r_Q), \qquad r_Q = O(1/M_Q)\ . \tag{6}$$

We note that although we use Eq.(6) to account for the fact that the annihilation of the $Q\overline{q}$ pair is "smeared" over the quantum-mechanical fluctuations in the position of the heavy quark, we do *not* intend to include finite M_Q effects when computing the light quark wavefunction ψ (in particular, we ignore the recoil and color-magnetic moment of the heavy quark). $\psi(\mathbf{r})$ is therefore independent of M_Q for all $\mathbf{r}$. Of course, a completely consistent treatment would include finite M_Q effects when computing $\psi(\mathbf{r})$ as well as when computing the overlap. However, we show in section V and in an Appendix that our neglect of these effects on $\psi(\mathbf{r})$ should either tend to *suppress* the results we are interested in, or should amount to neglecting a small correction to the wavefunction as computed with $M_Q = \infty$.

III. MODEL-INDEPENDENT ANALYSIS OF RELATIVISTIC AND SHORT-DISTANCE EFFECTS

Assuming that the heavy quark can be treated as a fixed, static source, we can obtain an *exact* result for the short-distance limit of the light quark wavefunction, because we know from perturbative QCD that at short-distances the light quark experiences a Coulomb potential modified by the running coupling constant

$$V_{\rm coul}(r) \quad \equiv \quad -\frac{\overline{\alpha}(r)}{r} \quad = \quad -\frac{4}{3}\frac{\alpha_s(r)}{r}\ . \tag{7a}$$

To leading-log order α_s is given by

$$\alpha_s(r) \quad = \quad -\frac{1}{4\pi b_0 \log(r\Lambda_{\rm pot})}\,, \tag{7b}$$

where b_0 is the one-loop β-function coefficient

$$b_0 \quad = \quad \frac{1}{8\pi^2}(11 - \tfrac{2}{3}N_f) \tag{7c}$$

and where $\Lambda_{\rm pot}$ is the scale mass appropriate to the $Q\overline{q}$ potential. $\Lambda_{\rm pot}$ must be related to the fundamental QCD scale parameter. In section IV we derive the connection between $\Lambda_{\rm pot}$ and $\Lambda_{\overline{\rm MS}}$, the QCD scale parameter in the modified-minimal-subtraction scheme.

The short-distance properties of the light quark wavefunction $\psi(\mathbf{r})$ can therefore be obtained from the solution to the Dirac equation

$$[\boldsymbol{\alpha}\cdot(-i\boldsymbol{\nabla}) \quad + \quad V_{\rm coul}(r) \quad + \quad \beta m]\,\psi_{\rm coul}(\mathbf{r}) \quad = \quad \xi\psi_{\rm coul}(\mathbf{r})\,, \tag{8}$$

where m is the *current* mass of the quark, and ξ is the energy eigenvalue of the stationary state $\psi(\mathbf{r},t) = e^{-i\xi t}\psi(\mathbf{r})$. Note that we have used $V_{\rm coul}$ as the zeroth component of a *vector* potential in Eq.(8), as given by perturbative QCD. We use the notation $\psi_{\rm coul}$ to emphasize that the wavefunction is calculated in the approximation that the $Q\overline{q}$ potential is a "pure" leading-log Coulomb potential. The form of the ground state solution to Eq.(8) is given in Eq.(3). Since the normalization $\mathcal{N}$ and energy eigenvalue ξ depend on the long-distance properties of the wavefunction, they cannot be determined from the short-distance potential $V_{\rm coul}$. However, the short-distance limit of $\psi(\mathbf{r})$ is actually independent of ξ (and m) [see Eq.(11)], and we will be able to draw significant conclusions about the behavior of the decay constant without knowing the wavefunction normalization (section V).

It is therefore of interest to study the Dirac equation with ξ and m set to zero, and without considering the overall normalization $\mathcal{N}$

$$\left[\boldsymbol{\alpha}\cdot(-i\boldsymbol{\nabla}) \quad - \quad \frac{\overline{\alpha}(r\Lambda_{\rm pot})}{r}\right]\begin{pmatrix} -i\chi_{\rm coul}(r)\,|_{\xi=0} \\ \boldsymbol{\sigma}\cdot\hat{\mathbf{r}}\,\phi_{\rm coul}(r)\,|_{\xi=0}\end{pmatrix} \quad = \quad 0 \tag{9}$$

[where we use the notation $|_{\xi=0}$ for quantities computed from the Dirac equation with ξ set to zero]. Note that Eq.(9) contains the *explicit* scale parameter $\Lambda_{\rm pot}$, which is connected to $\Lambda_{\overline{\rm MS}}$. The scale for the wavefunction $\psi_{\rm coul}(\mathbf{r})$ is therefore set explicitly by the QCD scale parameter, as expected of exact QCD.

To solve Eq.(9), we proceed in the usual way by separating it into two coupled, first order radial equations, which are identical in form to the usual set of radial Dirac equations, except that ξ is here set to zero. We express $\chi_{\rm coul}$ and $\phi_{\rm coul}$ as *dimensionless* functions of $r\Lambda_{\rm pot}$, and require the solution to be normalizable at the origin [the long-distance properties of the solution to Eq.(9) are not physically meaningful]. It turns out that the solution can be obtained using (asymptotic) series expansions in powers of the running coupling constant![24] We have

$$\chi_{\rm coul}(r\Lambda_{\rm pot})\,|_{\xi=0} \quad = \quad \sum_{n=0}^{\infty} g_n \left(-\frac{\alpha_0}{\log(r\Lambda_{\rm pot})}\right)^n , \tag{10a}$$

$$\phi_{\rm coul}(r\Lambda_{\rm pot})\,|_{\xi=0} \quad = \quad \sum_{n=0}^{\infty} f_n \left(-\frac{\alpha_0}{\log(r\Lambda_{\rm pot})}\right)^n , \tag{10b}$$

where for convenience we have defined the (positive) constant α_0 according to

$$\alpha_0 \equiv \frac{1}{3\pi b_0} . \tag{10c}$$

The coefficients in Eqs.(10a) and (10b) are obtained from the recursion relations

$$\begin{aligned} g_0 &\equiv 1, \quad f_0 = 0, \\ f_n &= \frac{1}{2}\left[\frac{(1-n)}{\alpha_0} f_{n-1} + g_{n-1}\right] , \\ g_n &= -\frac{\alpha_0}{n} f_n , \end{aligned} \tag{10d}$$

where for convenience we have normalized the coefficients with $g_0 \equiv 1$. Unfortunately, the series expansions in Eqs.(10a,b) only converge asymptotically; however, the first few terms in the series for $\chi_{\rm coul}\,|_{\xi=0}$ are sufficient to compute it to an accuracy of better than one percent for $r\Lambda_{\rm pot} \lesssim 0.1$ (the series for $\phi_{\rm coul}\,|_{\xi=0}$ is somewhat less convergent).

As described above, the wavefunction $(\chi_{\rm coul}\,|_{\xi=0},\ \phi_{\rm coul}\,|_{\xi=0})$ in the leading-log Coulomb potential $V_{\rm coul}(r)$ with ξ and m set to zero is an *exact* result for the light quark wavefunction in the short-distance limit, because in this limit the perturbative result, $V_{\rm coul}$, for the $Q\bar{q}$ potential is reliable and because the energy eigenvalue ξ and mass m can be neglected to leading order. Of particular interest to us is the upper wavefunction component. Denote the exact result by $\chi(r)$. Then we have

$$\chi(r) \quad \underset{r\to 0}{\longrightarrow} \quad \chi_{\rm coul}(r)\,|_{\xi=0} \quad + O\left[(\xi+m)\,r/\log(r\Lambda_{\rm pot})\right] . \tag{11}$$

We can therefore truncate the asymptotic series Eq.(10a) to obtain the following *exact* result for the leading short-distance behavior of the light quark wavefunction [assuming, as always, that the mass of the heavy quark is large enough to treat it as a fixed source]

$$\chi(r) \quad \underset{r\to 0}{\longrightarrow} \quad \chi(0)\left[1 - \left(-\frac{1}{18\pi^2 b_0^2 \log(r\Lambda_\chi)}\right) + O\left(\frac{1}{\log^3(r\Lambda_\chi)}\right)\right] , \tag{12a}$$

where $\chi(0)$ is finite [cf. $g_0 = 1$ in Eq.(10a)]. The scale mass Λ_χ differs from the scale mass Λ_{pot} appearing in the series expansion of Eq.(10a)

$$\Lambda_\chi = \Lambda_{\text{pot}}\, e^{-(\alpha_0^2+1)/4} , \tag{12b}$$

because the $O(1/\log^2)$ term in an expansion such as Eq.(10a) cannot be neglected with respect to the $O(1/\log)$ term; the former term gets absorbed into the scale mass appearing in the latter term, with a truly higher order remainder of $O(1/\log^3)$.

We stress that the Dirac wavefunction at the origin in the leading-log Coulomb potential is finite, unlike the more familiar case of the wavefunction in a Coulomb potential of fixed strength [such as the Dirac wavefunction of an electron in the field of a proton; $\chi_{\text{em}}(r \to 0) \sim r^{-\alpha_{\text{em}}^2/2}$ is singular]. This result does not seem to have been appreciated in the literature. The finiteness of $\chi(0)$ in the leading-log potential is evidently due to the fact that the running coupling constant $\overline{\alpha}(r)$ tends to zero sufficiently rapidly as $r \to 0$. [25] However, this result differs from the singular behavior found by L. Durand for a relativistic spin-0 equation in the same potential. [26]

Our analysis of the short-distance limit of $\psi(\mathbf{r})$ has important consequences for the decay constant. However, the applicability of the solution to Eq.(8) is somewhat limited because the perturbative result for the $Q\overline{q}$ potential, $V_{\text{coul}}(r)$, restricts our use of $\chi_{\text{coul}}(r)$ to very short distances, $r \ll \Lambda_{\text{pot}}^{-1}$. Note, in particular, that the solution to Eq.(8) will exhibit a form of Klein paradox as r approaches $\Lambda_{\text{pot}}^{-1}$, where the potential V_{coul} blows up, causing unphysical pair creation effects.

IV. THE $Q\overline{q}$ SYSTEM IN AN EFFECTIVE ACTION MODEL OF QCD

As outlined in the Introduction, we have found that an effective action model (EAM) of QCD may provide us with a reliable model of the $Q\overline{q}$ system at essentially *all* length scales. In this section, we only summarize the essential features of this model, and show how it can be applied to the $Q\overline{q}$ system in order to calculate the light quark wavefunction. [27] A full account of our application of the effective action model to $Q\overline{q}$ mesons (and Qqq baryons) will be published elsewhere. [19]

The effective action model (EAM) we use is defined by an effective Lagrangian density for the gauge fields, $\mathcal{L}_{\text{eff}}^{\text{gauge}}$, which includes leading renormalization-group corrections to the gauge-field Lagrangian in classical chromodynamics [15,18]

$$\mathcal{L}_{\text{eff}}^{\text{gauge}}(\mu, -\tfrac{1}{2}F^{a\mu\nu}F^a_{\mu\nu}) = -\tfrac{1}{4}\overline{g}^{-2}(\mu, -\tfrac{1}{2}F^{a\mu\nu}F^a_{\mu\nu})\, F^{a\mu\nu}F^a_{\mu\nu} , \tag{13a}$$

where the running coupling constant $\overline{g}(\mu, -\frac{1}{2}F^{a\mu\nu}F^{a}_{\mu\nu})$ here runs with the *field strength*. μ is the subtraction point. To leading-logarithm order $\overline{g}$ is given by

$$\overline{g}^{-2}(\mu, -\tfrac{1}{2}F^{a\mu\nu}F^{a}_{\mu\nu}) = g^{-2}\left[1 + \tfrac{1}{4}b_0 g^2 \log\left(\frac{-\frac{1}{2}F^{a\mu\nu}F^{a}_{\mu\nu}}{\mu^4}\right)\right], \tag{13b}$$

where $g \equiv \overline{g}(\mu, \mu^4)$ and b_0 is the one-loop β-function coefficient [see Eq.(7c)]. We note that Eqs.(13a,b) are essentially exact results for the gauge-field effective Lagrangian in the limit of infinitely large, space-time independent field strengths.

The total effective Langrangian density $\mathcal{L}^{Q\overline{q}}$ for the $Q\overline{q}$ system is the sum of the effective Lagrangian for the gauge fields plus a fermionic Lagrangian $\mathcal{L}^{F}$

$$\mathcal{L}^{Q\overline{q}} \quad = \quad \mathcal{L}^{\text{gauge}}_{\text{eff}} + \mathcal{L}^{F}, \tag{13c}$$

where $\mathcal{L}^{F}$ has the usual form (ignoring the kinetic term for the heavy quark, which we will treat as an infinitely heavy point source) [18]

$$\mathcal{L}^{F} = i\overline{\psi}\gamma^{\mu}(\partial_{\mu} - i\tfrac{1}{2}\lambda^{a}A^{a}_{\mu})\psi \; + \; \overline{\psi}_{M}\gamma^{\mu}\tfrac{1}{2}\lambda^{a}A^{a}_{\mu}\psi_{M} \; - \; m\overline{\psi}\psi \; - \; M'\overline{\psi}_{M}\psi_{M} \,. \tag{13d}$$

ψ_M is the Dirac wavefunction for the heavy quark of (bare) mass M', and ψ is the wavefunction for the light quark of *current* mass m.

Following the approach used in our previous work [18] (and used by Adler and Piran for $Q\overline{Q}$ systems [15–17]), we assume that the action corresponding to $\mathcal{L}^{Q\overline{q}}$ is minimized by an *Abelian* configuration of fields and currents. This amounts to dropping the color indices in Eqs.(13), and assigning equal and opposite Abelian charges of magnitude $\sqrt{4/3}$ to the quarks [in order to satisfy the $SU(3)_C$ Casimirs for a color singlet $Q\overline{q}$ state]. The classical equations of motion which minimize the resulting Abelian action are similar to the usual Abelian Maxwell equations for the gauge fields, plus an Abelian Dirac equation for the light quark (the heavy quark is treated as an external source). However, the fields and sources appear to be embedded in a "medium" whose response to applied fields is described by a *nonlinear* dielectric ϵ, which is a function of the field strengths

$$\epsilon(F) \quad = \quad \frac{\partial \mathcal{L}^{\text{gauge}}_{\text{eff}}(F)}{\partial\left(\frac{1}{2}F\right)}, \qquad F(\mathbf{r}) = \mathbf{E}^2 - \mathbf{B}^2 \,. \tag{14a}$$

In the leading-log model for the gauge-field effective Lagrangian, defined by Eqs.(13a,b), $\epsilon(F)$ is given by

$$\epsilon(F) = \tfrac{1}{4}b_0 \log\left(\frac{F}{\kappa^2}\right) = \tfrac{1}{4}b_0 \log\left(\frac{\mathbf{E}^2 - \mathbf{B}^2}{\kappa^2}\right), \tag{14b}$$

where the scale parameter $\kappa^{1/2}$ is related to the subtraction point μ, $\kappa^2 \equiv \mu^4/e \times \exp[-4/(b_0 g^2)]$.

Because of the nonlinear dielectric, the Abelian Maxwell and Dirac equations are highly coupled and nonlinear.[18] To simplify the problem of solving these equations, we here make one final approximation, which is to ignore the 3-vector part $\mathbf{j}(\mathbf{r})$ of the light quark Abelian current density $j^\mu(\mathbf{r})$

$$j^\mu(\mathbf{r}) \quad = \quad \sqrt{\frac{4}{3}}\bar{\psi}(\mathbf{r})\gamma^\mu\psi(\mathbf{r})$$

compared to the charge density $j^0(\mathbf{r})$. One can show quite generally that this approximation is *exact* close to the point charge of the heavy quark (which is the region where we wish to evaluate $\psi(\mathbf{r})$ in order to calculate the decay constant). This approximation will be improved in future work.[19] With $\mathbf{j}(\mathbf{r}) \equiv 0$, the Maxwell equations for the (Abelianized) color-magnetic field are satisfied by $\mathbf{B} = 0$.

We thus arrive at the following set of equations of motion. We have a set of nonlinear Maxwell equations for the (Abelianized) color-electric field $\mathbf{E}$

$$\nabla \cdot \left(\epsilon(E^2)\mathbf{E}\right) \quad = \quad \sqrt{\frac{4}{3}}\left[\psi^\dagger(\mathbf{r})\psi(\mathbf{r}) - \delta^3(\mathbf{r})\right] , \tag{15a}$$

$$\mathbf{E}(\mathbf{r}) \quad \equiv \quad -\nabla A^0(\mathbf{r}) , \tag{15b}$$

where we have written the charge density in the $Q\bar{q}$ center-of-mass frame, with the heavy quark described as a static point charge.

The Maxwell equations for $\mathbf{E}$ are coupled (*nonlinearly*) through the light quark source term and the vector potential A^0 to the Dirac equation for the light quark wavefunction

$$\left[\alpha\cdot(-i\nabla) \quad + \quad \sqrt{\frac{4}{3}}A^0(\mathbf{r}) \quad + \quad \beta m\right]\psi(\mathbf{r}) \quad = \quad \xi\psi(\mathbf{r}) , \tag{16}$$

The form of the ground state solution to Eq.(16) is given in Eq.(3).

In order to uniquely determine the solution to Eq.(16) we must impose suitable boundary conditions on $\psi(\mathbf{r})$. Unfortunately, we have been unable to find a solution without imposing (as an intermediate step) a baglike boundary condition, which breaks chiral symmetry explicitly. We take the boundary condition in the following familiar form

$$\left[\bar{\psi}(\mathbf{r})\psi(\mathbf{r})\right]_{r=R} \quad = \quad 0 . \tag{17a}$$

The radius R at which the boundary condition is imposed is determined by the requirement that the fermionic part of the energy-momentum tensor be conserved across the surface $r = R$. In the usual bag model,[22] the sources are confined by adding an external pressure term to the action, and the continuity of the energy-momentum tensor is equivalent to a condition of "pressure balance" between the pressure generated by the light quark on the bag surface and the external pressure. In the EAM, however, confinement of the $Q\bar{q}$ system arises *automatically* as a consequence of the nonlinear properties of the dielectric function ϵ (see below). In particular, *we do not impose an external bag pressure.* Thus, the bag radius R is determined in the EAM by requiring the light quark pressure to *vanish* on the bag surface, which results in the boundary condition

$$\frac{\partial}{\partial r}\left[\bar{\psi}(\mathbf{r})\psi(\mathbf{r})\right]_{r=R} \quad = \quad 0\ . \tag{17b}$$

Finally, we require that the light quark wavefunction be normalizable, with

$$\int_{r\leq R} d^3\mathbf{r}\,\psi^{\dagger}(\mathbf{r})\psi(\mathbf{r}) \quad = \quad 1\ . \tag{17c}$$

The EAM has one of the most important features expected of exact QCD, which is that it automatically interpolates between asymptotic freedom at short distances and linear confinement at long distances.[15] With respect to the $Q\bar{q}$ system, these properties are most clearly illustrated by considering the $Q\bar{q}$ vector potential, V_{eam} [cf. Eq.(16)]

$$V_{\text{eam}}(r) \quad \equiv \quad \sqrt{\frac{4}{3}}\,A^0(r)\ . \tag{18a}$$

The model exhibits asymptotic freedom at short-distances due to the properties of the dielectric function $\epsilon(E)$ in strong fields

$$V_{\text{eam}}(r) \underset{r\Lambda_{\text{pot}}\ll 1}{\longrightarrow} -\frac{1}{3\pi b_0\log(r\Lambda_{\text{pot}})}\ , \tag{18b}$$

in agreement with perturbative QCD [cf. Eq.(7)]. We derive the scale mass Λ_{pot} in terms of $\kappa^{1/2}$, $\Lambda_{\text{pot}} = e[\sqrt{3}\pi b_0\kappa]^{1/2}$. Adler and Piran have determined the connection between $\kappa^{1/2}$ and the QCD scale parameter $\Lambda_{\overline{\text{MS}}}$ from the quarkonium ($Q\overline{Q}$) potential. For three light quark flavors, $\Lambda_{\overline{\text{MS}}} = 0.959\kappa^{1/2}$ [for arbitrary N_f, see Ref.17], and we have

$$\Lambda_{\text{pot}} \quad = \quad 2.23\,\Lambda_{\overline{\text{MS}}} \qquad [N_f = 3]\ . \tag{18c}$$

This is close to the analogous result for scale mass in the quarkonium ($Q\overline{Q}$) potential, which equals[28] $2.63\,\Lambda_{\overline{\text{MS}}}$ [$N_f = 3$]. We conclude from Eqs.(18a,b) that the wavefunction in the EAM has the same leading short-distance behavior found in section III in the pure leading-log Coulomb potential [see Eqs.(10)–(12)].

The EAM exhibits linear-confinement because $\epsilon(E)$ has a zero in weak fields, $\epsilon(E = E_0 = 1.0\kappa) = 0$. [29] The total energy of the $Q\overline{q}$ system rises linearly for large bag radii R, and the vector potential $V_{\mathrm{eam}}(r)$ becomes linear as $r \to R$

$$V_{\mathrm{eam}}(r) \underset{r\to R}{\longrightarrow} \sigma r + \text{constant}, \qquad \sigma = (1.12\,\Lambda_{\overline{\mathrm{MS}}})^2 \quad [N_f = 3] \ . \tag{18d}$$

We note that although the *vector* potential V_{eam} rises linearly towards the bag edge, the light quark wavefunction $\psi(\mathbf{r})$ only exhibits the Klein paradox to a *negligible* extent. This is because the baglike boundary condition, Eq.(17a), acts as a partial infrared cut-off. [30]

The system of coupled nonlinear Maxwell and Dirac equations [Eqs. (15a,b) and (16)], subject to the boundary conditions Eqs.(17), are solved numerically. For $m = 0$ the bag radius which satisfies Eq.(17b) is $R = 1.21\kappa^{-1/2}$. The wavefunction components $\chi(r)$ and $\phi(r)$ are plotted in Figure 1, where the effect of the running coupling constant at short distances is clearly demonstrated: note especially that for constant $\overline{\alpha}$, we would have had $\phi(r \to 0) \propto \overline{\alpha}\chi(r \to 0) \to \infty$; instead we have $\chi \to$ finite and $\phi \to 0$, as we found in Eq.(10). [25]

V. RESULTS

A. Comparison of f_P in the leading-log potential and the EAM

Although the leading-log Coulomb potential gives us an exact, *model-independent* result for the light quark wavefunction at short-distances, its application to the decay constants must be made with care. According to Eq.(6) we must evaluate $\chi(r)$ at a distance r_Q of $O(1/M_Q)$ to get the decay constant, and even for r_Q corresponding to a meson as heavy as the D, the perturbative result for V_{coul} is unreliable, because of the large scale mass [see Eqs.(7) and (18c); assuming $M_c \approx 1.6$ GeV and $\Lambda_{\overline{\mathrm{MS}}} \approx 220$ MeV, we have $r_Q \approx .3\,\Lambda_{\mathrm{pot}}^{-1}$ and $\overline{\alpha}(r_Q) \approx 0.7$]. On the other hand, the EAM gives us a (hopefully) reliable model of the light quark wavefunction over the full extent of the meson [at the very least, the EAM and the leading-log Coulomb potential give the *same* result for the leading short-distance behavior, Eq.(12)]. By comparing the results for f_{rel} in the leading-log Coulomb potential and the EAM, we may be able to estimate the importance of the detailed long-distance behavior of the wavefunction.

To make specific calculations, we must give a definite connection between the overlap radius r_Q and the heavy quark mass M_Q. On physical grounds, we do not expect r_Q to be much smaller than $1/M_Q$; we make the convenient choice $r_Q = 1/M_Q$. Furthermore, since we do not know the connection between the quark mass M_Q and the meson mass M_P without a complete model calculation, we make the additional approximation $M_Q \approx M_P$

(which should hold in the limit of sufficiently large M_Q). Our choice for r_Q is therefore $r_Q \equiv 1/M_P$. Finally, since we have neglected finite M_Q effects in computing the wavefunction, the normalization $\mathcal{N}$ does not affect the M_Q dependence of $f_{\rm rel}$, and can be disregarded for our purposes. We show below that these approximations should only *suppress* the effects we are interested in.

We therefore compare the M_P dependence of the following *dimensionless* functions

$$\widetilde{f}_{\rm eam}(\widetilde{M}) \quad \equiv \quad \left[\widetilde{M}\right]^{-1/2} \chi_{\rm eam}(1/\widetilde{M}), \qquad [m=0] \tag{19}$$

and

$$\widetilde{f}_{\rm coul}(\widetilde{M})\,|_{\xi=0} \quad \equiv \quad \left[\widetilde{M}\right]^{-1/2} \chi_{\rm coul}(1/\widetilde{M})\,|_{\xi=0}\,, \qquad [m=0] \tag{20}$$

where

$$\widetilde{M} \quad \equiv \quad M_P \;/\; \Lambda_{\overline{\rm MS}} \tag{21}$$

and where we take $N_f = 3$ in the rest of this paper. We use $\chi_{\rm eam}$ and $\chi_{\rm coul}$ to denote the wavefunction as computed in the EAM and the leading-log Coulomb potential, respectively. The notation $|_{\xi=0}$ emphasizes that $\chi_{\rm coul}\,|_{\xi=0}$ was computed with the energy eigenvalue ξ set to zero.

We note that $\widetilde{f}_{\rm eam}(M_P/\Lambda_{\overline{\rm MS}})$ is *equal* to the actual decay constant $f_{\rm rel}(M_Q)$ [Eq.(6)] as computed in the EAM, up to an overall normalization [provided that $M_Q \approx M_P$, as discussed above Eq.(19)]. We stress that the scale for the decay constant in our calculations is *explicitly* set by the QCD scale parameter, as expected of exact QCD.

Note that we have defined the functions $\widetilde{f}$ with the light quark mass set to zero, which is necessary in order to be consistent with the condition $\xi = 0$ imposed on $\chi_{\rm coul}$ [see Eq.(11)]. This is of no consequence to the present line of inquiry, as we are interested in the M_Q dependence of the decay constants. Moreover, the u and d quark masses are very small, $m_u, m_d \ll \Lambda_{\overline{\rm MS}}$. A comparison of the decay constants for families of $Q\overline{q}$ mesons containing different flavors of light quark can be made in a more complete calculation in the EAM (to be pursued in future work).

To compare the evolution of these functions with increasing meson mass, it is convenient to normalize them to have the same value at some small mass. We choose this "reference" mass near the smallest mass at which we may trust the wavefunction in the leading-log Coulomb potential [31]

$$\widetilde{f}_{\rm eam}(\widetilde{M}_{\rm ref}) \quad \equiv \quad \widetilde{f}_{\rm coul}(\widetilde{M}_{\rm ref})\,|_{\xi=0} \quad \equiv \quad 1, \qquad \widetilde{M}_{\rm ref} \equiv 3.5\,. \tag{22}$$

$\widetilde{f}_{\rm eam}$ and $\widetilde{f}_{\rm coul}\,|_{\xi=0}$ are plotted in Figure 2 over a continuous range of values of $\widetilde{M}$. We see that the curves are very similar over the whole range, all the way down to the "reference" mass.[32] The outstanding feature of both curves is that they possess a maximum. This result appears to be due entirely to the running of the QCD coupling constant, plus relativistic dynamics for the light quark. These effects cause $\chi(r = 1/M_P)$ to *increase* as M_P increases, which "temporarily" overcomes the decrease in the kinematical factor $\sqrt{1/M_P}$. Thus, $\tilde{f}$ rises with M_P before it falls.[25] The location of the maxima in the two curves differs only by $\approx 5\%$. The only significant difference between the two functions is that the maximum of $\tilde{f}_{\rm coul}|_{\xi=0}$ is somewhat more pronounced. This is due to the increasing effects of the Klein paradox on $\chi_{\rm coul}(1/M_P)$ as $M_P \to \Lambda_{\rm pot}$ [from above], which cause $\chi_{\rm coul}$ to change concavity (forcing $\chi_{\rm coul} > \chi_{\rm eam}$) and then to decrease extremely rapidly, as if approaching an asymptote at $M_P = \Lambda_{\rm pot}$.

From this analysis, we conclude that the EAM gives an essentially model-independent account of relativistic and short-distance QCD effects on f_P. Furthermore, the long-distance properties of the wavefunction apparently play a relatively unimportant role in determining the main features of the M_Q dependence of $f_P(M_Q)$, even at meson masses as low as $M_P \approx$ 3–4$\Lambda_{\overline{\rm MS}}$.

B. Comparison with non-relativistic quark model

We briefly comment on the relationship between our relativistic decay constant $f_{\rm rel}$ [Eq.(7)] and the constant as calculated in non-relativistic quark models.[1-3] In these models, the decay constant is given by

$$f_{\rm NR}(M_Q) \quad = \quad \sqrt{\frac{12}{M_P}}\,\Psi_{\rm NR}(0)\ , \tag{23}$$

where the non-relativistic wavefunction $\Psi_{\rm NR}(0)$ depends on the reduced mass of the $Q\bar{q}$ pair [in this picture, one generally assigns a *constituent* mass to the light quarks (u,d,s) of a few hundred MeV]. Several authors [e.g. Refs.2,3] claim that $\Psi_{\rm NR}$ is already independent of M_Q at meson masses as small as $M_P \approx M_K$. In this extreme or "naive" picture, the $1/\sqrt{M_P}$ factor in Eq.(23) carries the full mass dependence of the decay constant for $M_P \gtrsim M_K$. For comparison with our results, we plot the function $[\widetilde{M}_{\rm ref}/\widetilde{M}\,]^{1/2}$ in Figure 2, which is the analogue to our $\tilde{f}_{\rm eam}$ in the "naive" non-relativistic picture. Note that we have normalized the "naive" prediction to 1 at the reference mass of Eq.(22). From Figure 2 we clearly see that *if* the "naive" picture correctly predicts the value of f_P for a light meson, such as the K [as claimed in Refs.2,3], then this picture will significantly under-estimate the value of f_P for heavier states.

We stress that whatever behavior is found in non-relativistic (and quasi-relativistic[4]) quark models, they do not include the significant relativistic effects that we have found using the Dirac equation.

C. Quantitative predictions

We now use the EAM to extract some quantitative predictions for the decay constants. First, we note again that relativistic plus short-distance QCD effects are enough to cause $f_{\rm rel}$ to have a maximum. From $\widetilde{f}_{\rm eam}$ [Eq.(19)] we can place a *lower* bound [33] on the *meson mass* $M_{\rm peak}$ at which $f_{\rm rel}$ is a maximum

$$M_{\rm peak} \quad \gtrsim \quad 4.83\,\Lambda_{\overline{\rm MS}}\,. \tag{24}$$

With the reasonable choice $\Lambda_{\overline{\rm MS}} \approx 220$ MeV [which Adler and Piran obtain [17] in their fit to the quarkonium potential], we have $M_{\rm peak} \gtrsim 1.1$ GeV, which is *at least half-way from the K to the D.* [34]

In the same way, we can use our result for $\widetilde{f}_{\rm eam}$ to place a lower bound [33] on the ratio between the decay constants of the B and D mesons

$$\frac{f_B}{f_D} \quad \gtrsim \quad \sqrt{\frac{M_D}{M_B}} \times 1.26 \qquad [\Lambda_{\overline{\rm MS}} = 220 \text{ MeV}]\,, \tag{25}$$

where we have written our result in terms of a correction to the "naive" non-relativistic prediction, which shows how significant the relativistic and short-distance effects are. As we describe in a footnote, [33] these effects have been underestimated by using the Compton wavelength of the meson, instead of the heavy quark, to evaluate the $Q\overline{q}$ overlap [compare Eq.(6) with Eqs.(19) and (21)]. We estimate the additional enhancement to f_B/f_D, when using $r_Q = 1/M_Q$, to be ≈ 1.05 [assuming M_c =1.6 GeV and M_b =5.0 GeV]. Thus, the relativistic and short-distance effects amount to a correction to the "naive" ratio of *at least* 1.32 [this is still a lower bound, because r_Q may be larger than $1/M_Q$ [33]].

We can also estimate the effect on $f_{\rm rel}$ of having ignored finite M_Q effects on $\psi(\mathbf{r})$. [35] We divide these effects into two groups. The first group accounts for effects due to the recoil and color-magnetic moment of the heavy quark, and the second group accounts for the effect of quantum-mechanical fluctuations in the position of the heavy quark, on the short-distance $Q\overline{q}$ potential $V_{\rm coul}$ [this effect is discussed in the Appendix, and is found to be small]. If we include the effects of the first group as first-order perturbations on the wavefunction with $M_Q = \infty$ we expect that the pseudoscalar bag radii will be reduced (by the dominant attractive magnetic moment term). The wavefunction normalization $\mathcal{N}$ should then favor mesons with heavier Q quarks. As a rather conservative estimate of this effect we take

$$\frac{\mathcal{N}_B}{\mathcal{N}_D} \approx \left[\frac{R_D}{R_B}\right]^{3/2} \approx (1.10)^{3/2}$$

which gives an additional enhancement of $\approx$ 1.15 to Eq.(25).

We therefore obtain the following lower limit for ratio of the decay constants, which includes short-distance, relativistic and finite M_Q effects

$$\frac{f_B}{f_D} \gtrsim \sqrt{\frac{M_D}{M_B}} \times 1.45 \sim 0.86 \qquad [\Lambda_{\overline{\rm MS}} = 220 \text{ MeV}] , \tag{26}$$

As one final application of our results, we consider the ratio of the D to K decay constants. Of course, our methods of treating finite M_Q effects are of questionable validity when applied to the s-quark [although for the application to the K meson, we would assign a *constituent* mass to the s-quark].[32] Furthermore, our model explicitly breaks chiral symmetry, which could be very important for the K. However, these objections are no more severe than for other models, and so we try it anyway. Our calculation of $\tilde{f}_{\rm eam}$ gives the lower bound

$$\frac{f_D}{f_K} \gtrsim \sqrt{\frac{M_K}{M_D}} \times 2.00 \sim 1 \qquad [\Lambda_{\overline{\rm MS}} = 220 \text{ MeV}] , \tag{27}$$

where the correction to the "naive" non-relativistic result is entirely due to relativistic and "short"-distance effects.

VI. SUMMARY AND OUTLOOK

We have found that relativistic and short-distance QCD effects on the pseudoscalar decay constant f_P are very significant. These effects are completely missed by non-relativistic and quasi-relativistic potential models. These effects cause f_P to have a maximum at a meson mass of *at least* 1.1 GeV, half-way from the K to the D. Our results indicate that the ratio f_B/f_D is significantly closer to 1 than indicated by the "naive" non-relativistic picture $f_B/f_D \approx (M_D/M_B)^{1/2} \approx 0.6$. We also made a rough estimate for $f_D/f_K \approx 1$, although the reliability of our methods is difficult to judge when applied to the K system.

We note that Shifman and Voloshin have recently put forward another effect[20] which would further contribute to the trend exhibited by our results. Their effect is due to a renormalization of the matrix element which defines f_P [Eq.(1)], from the large momentum scale $\sim M_Q$ at which the matrix element is defined, down to the soft momentum scale $\sim R^{-1}$ which characterizes the $Q\bar{q}$ wavefunction. However, this effect is far less significant than the effects we have considered here. For example, their correction to f_B/f_D amounts to a factor $\approx$ 1.09 [at $\Lambda_{\overline{\rm MS}}$ =220 MeV], while our effects give a correction of at least 1.32 [see discussion below Eq.(25)].

We note that we can easily extend the techniques of this paper to get bounds on the so-called B-parameter, which enters the calculation of important processes such as B_d–$\overline{B}_d$ mixing, etc.

The results presented here can also be made more complete by performing a comprehensive calculation of the properties of the $Q\overline{q}$ system in the effective action model [e.g. this would provide us with the wavefunction normalization as a function of M_Q]. However, in order to make definitive predictions for the decay constants, we must fix the connection between the overlap distance r_Q and the heavy quark mass M_Q. Unfortunately, there does not appear to be a way to do this from first principles, so we would have to introduce an adjustable parameter, i.e. $r_Q = \lambda\, M_Q^{-1}$, and the only available experimental result we could use to fit λ is the value of f_K. We feel that this calculation is worth pursuing.

We are working on these problems at present.

ACKNOWLEDGEMENT

We thank B. Margolis for fruitful discussions during the preliminary stages of this work.

APPENDIX

In this Appendix, we justify our two step procedure for including finite M_Q effects on $f_{\mathrm{rel}}(M_Q)$. Recall that we have *not* included these effects when calculating the light quark wavefunction $\psi(\mathbf{r})$, since we have "fixed" the heavy quark at the origin in order to write down the Dirac equation in a static, central potential [i.e. Eqs.(8) and (16)]. However, we *do* include finite M_Q effects when computing the overlap of the $Q\overline{q}$ pair [according to Eq.(6)].

As described below Eq.(25), we divide finite M_Q effects on $\psi(\mathbf{r})$ into two classes. Here we consider the effect due to quantum-mechanical fluctuations in the position of the heavy quark, on the short-distance $Q\overline{q}$ potential V_{coul} [Eq.(7)]. We make the reasonable assumption that these fluctuations would have the effect of "softening" $V_{\mathrm{coul}}(r)$ within the distance r_Q, which is the "spread" in the heavy quark wavefunction used in Eq.(5) to evaluate the $Q\overline{q}$ overlap. For the sake of illustration, we suppose that the "softening" can be parametrized in the following form [cf. Eq.(7)]

$$V_{Q\overline{q}}(r) \underset{r\to 0}{\longrightarrow} \begin{cases} V_{\mathrm{coul}}(r), & r_Q \le r \ll \Lambda_{\mathrm{pot}}^{-1} \\ V_{\mathrm{soft}}(r), & r \le r_Q\ , \end{cases} \tag{A1}$$

where we assume that V_{soft} is bounded

$$|V_{\mathrm{soft}}(r \le r_Q)| \ \le\ \text{constant} \ =\ |V_{\mathrm{coul}}(r_Q)| \tag{A2}$$

as suggested by Gauss' Law, if the heavy quark was "smeared" over the distance r_Q and if $\overline{\alpha}(r)$ was constant.

The light quark wavefunction in the potential of Eq.(A1) is obtained by dividing the solution in the two regions $r > r_Q$ and $r < r_Q$. We concentrate on the upper component

$$\chi(r) \underset{r\to 0}{\longrightarrow} \begin{cases} \chi_{\mathrm{coul}}(r), & r_Q \le r \ll \Lambda_{\mathrm{pot}}^{-1} \\ \chi_{\mathrm{soft}}(r), & r \le r_Q \ , \end{cases} \tag{A3}$$

Note that $\chi(r > r_Q)$ is equal to the wavefunction in the "pure" leading-log Coulomb potential, $\chi_{\mathrm{coul}}(r)$, and is therefore *independent* of V_{soft} [for sufficiently small $r > r_Q$ it is given by the expression in Eq.(12a), up to an overall normalization]. The wavefunctions in the two regions are matched by requiring continuity at $r = r_Q$.

Eqs.(A1)–(A3) should hopefully form a reasonable parametrization of the effects of the "softening" of the leading-log Coulomb potential. On the other hand, we find that the leading behavior of $\chi_{\mathrm{soft}}(r)$ is actually *independent* of the details of V_{soft}, as long as it is bounded. We find:

$$\chi_{\mathrm{soft}}(r) \quad = \quad \chi_{\mathrm{soft}}(0)\left[1 - \tfrac{1}{6}\xi'^2 r^2 \ + \ O(\xi'^4 r^4)\right], \qquad r \le r_Q, \tag{A4}$$

where

$$\xi' \equiv \xi - V_{\mathrm{coul}}(r_Q) \ . \tag{A5}$$

Note that the leading r-dependent term in $\chi_{\mathrm{soft}}(r)$ is *independent* of the details of V_{soft}, as claimed. The next-to-leading r-dependent term is *also* independent of V_{soft}. We have not shown next-to-leading terms that do depend on the details of V_{soft}; these are smaller than the $O(r^4)$ term given, unless $r \ll r_Q$, and so are negligible.

We see from Eqs.(A4) and (A5) that the "softened" wavefunction is essentially constant over $r < r_Q$

$$\chi_{\mathrm{soft}}(r_Q) \ = \ \chi_{\mathrm{soft}}(0)\left[1 - \tfrac{1}{6}\overline{\alpha}^2(r_Q) \ + \ O\left(\overline{\alpha}^4(r_Q)\right)\right] \tag{A6}$$

where we have neglected $\xi \ = \ O(\Lambda_{\overline{\mathrm{MS}}})$ compared to $V_{\mathrm{coul}}(r_Q) \ = \ O(\overline{\alpha}\, M_Q)$ [note that it is essential to account for the r_Q dependence of $V_{\mathrm{coul}}(r_Q)$ in Eq.(A4), in order to obtain the correct r_Q dependence of $\chi_{\mathrm{soft}}(r_Q)$].

We conclude that if we average over the $Q\overline{q}$ overlap by using the light quark wavefunction in a "soft" $Q\overline{q}$ potential [in Eq.(5)], we expect to obtain a *small* correction to Eq.(6) for f_{rel}. Note that Eq.(A6) can only be used to estimate this correction when $\overline{\alpha}(r_Q)$ is small, so that perturbation theory for V_{coul} is reliable. However, since the "true" $Q\overline{q}$ potential ought to be less singular than V_{coul} when $\overline{\alpha}$ gets large, we *still* expect $\chi_{\mathrm{soft}}(r)$ to be essentially constant and the correction to Eq.(6) for f_{rel} to be small.

REFERENCES AND FOOTNOTES

[1] A review of several potential model calculations of f_P can be found in S. Godfrey, Phys. Rev. **D33**, 1391 (1986).

[2] M. Suzuki, Phys. Lett. **162B**, 392 (1985).

[3] S.N. Sinha, Phys. Lett. **178B**, 110 (1986).

[4] S. Godfrey and N. Isgur, Phys. Rev. **D32**, 189 (1985).

[5] L.J. Reinders, Bonn University Report No. BONN-HE-88-04, 1988.

[6] S. Narison, Phys. Lett. **198B**, 104 (1987).

[7] C.A. Dominguez and N. Paver, Phys. Lett. **197B**, 423 (1987); **199B**, 596(E) (1987).

[8] M.B. Gavela *et al.*, Phys. Lett. **206B**, 113 (1988).

[9] T.A. DeGrand and R.D. Loft, University of Colorado Report No. COLO-HEP-165, 1987.

[10] R.M. Woloshyn *et al.*, TRIUMF Report No. TRI-PP-87-62, 1987.

[11] D. Izatt, C. DeTar and M. Stephenson, Nucl. Phys. **B199**, 269 (1982).

[12] Mark III Collaboration, J. Adler *et al.*, Phys. Rev. Lett. **60**, 1375 (1988).

[13] G. Matinyan and G.K. Savvidy, Nucl. Phys. **B134**, 539 (1978).

[14] H. Pagels and E. Tomboulis, Nucl. Phys. **B143**, 485 (1978).

[15] S.L. Adler, Phys. Rev. **D23**, 2905 (1981); **D24**, 1063(E) (1981); Phys. Lett. **110B**, 302 (1982); Nucl. Phys. **B217**, 381 (1983).

[16] S.L. Adler and T. Piran, Rev. Mod. Phys. **56**, 1 (1984); Phys. Lett. **113B**, 405 (1982).

[17] S.L. Adler and T. Piran, Phys. Lett. **117B**, 91 (1982).

[18] B. Margolis, R.R. Mendel and H.D. Trottier, Phys. Rev. **D33**, 2666 (1986); and in *Proceedings of the XXIII International Conference on High Energy Physics*, 16–23 July, 1986, Berkeley, California (World Scientific, Singapore, 1987). The effective action model was originally applied to the $Q\overline{q}$ system by B. Margolis and R.R. Mendel in Phys. Rev. **D30**, 621 (1984).

[19] B. Margolis, R.R. Mendel and H.D. Trottier, in preparation.

[20] M.B. Voloshin and M.A. Shifman, Sov. J. Nucl. Phys. **45**, 292 (1987) [Yad. Fiz. **45**, 463 (1987)].

[21] See, for example, E.D. Commins and P.H. Bucksbaum, *Weak Interactions of Leptons and Quarks* (Cambridge Univ. Press. Cambridge, UK, 1983).

[22] See, for example, Ref.11.

[23] Eq.(5) can be derived using the methods of J.F. Donoghue and K. Johnson, Phys. Rev. **D21**, 1975 (1980).

[24] This is very different from the usual series expansion for the Dirac equation in a Coulomb potential of fixed strength; in that case, the expansion is in powers of r, with (constant) α appearing in the series coefficients [see, for example, Ref.18].

[25] The fact that $\chi(r)$ rises extremely rapidly towards a finite value as $r \to 0$ is, in large measure, responsible for the importance of the relativistic and short-distance QCD effects on $f_{\rm rel}$. It might be argued that this rapid rise is some kind of a "residual" effect of an unphysical singularity that would be present if $\overline{\alpha}(r=0)$ was non-zero. To show that this is not the case, we suppose that $\overline{\alpha}(r \to 0)$ decreases to a small value, α. Then $\chi(r \to 0) \sim (\Lambda_{\overline{\rm MS}}\, r)^{-\alpha^2/2}$ [$\Lambda_{\overline{\rm MS}}$ is the only available scale]. The singularity becomes important only at a distance $r\Lambda_{\overline{\rm MS}} \approx \alpha^{-1}e^{-2/\alpha^2}$, where the wavefunction exceeds unity [see, for example, J.D. Bjorken and S.D. Drell, *Relativistic Quantum Mechanics* (McGraw Hill, New York, 1964)]. Even if we take α as large as 0.5, this corresponds to a very short distance, $r\Lambda_{\overline{\rm MS}} \approx 10^{-3}$, which is much smaller than the radius $r_Q \approx 1/M_Q$ at which we evaluate the wavefunction to obtain the decay constants of the D and B mesons, for example. Thus, we conclude that the rapid rise of the light quark wavefunction in the leading-log QCD potential is a physical effect.

[26] L. Durand, Phys. Rev. **D32**, 1257 (1985). The spin-0 equation with running α_s is used to calculate f_P in P. Cea

[27] The application of the EAM to the $Q\overline{q}$ system described here is an extension of some earlier work by us, in collaboration with B.Margolis (see Ref.18). We have previously shown that the EAM should give a good quantitative account of the properties of mesons and baryons with one heavy quark. However, the methods outlined here represent a significant improvement over our earlier treatment [in particular, in our earlier work we replaced the running coupling constant in the Dirac equation by a *fixed* "effective" coupling; here we succeed in including the full running coupling constant in the Dirac equation].

[28] See Ref.17, and references therein.

[29] For example, if we consider the infrared energy of a single isolated charge, we find that $\epsilon E \propto 1/r^2$ due to flux conservation, while $E \to E_0$ as $r \to \infty$. Thus the energy density DE goes like $1/r^2$, while the volume element goes like r^3, and the infrared energy of the isolated charge diverges linearly [Ref.14].

[30] This is true only for bag radii R near the radius which satisfies the energy-momentum continuity condition, Eq.(17b), $R \approx \kappa^{-1/2}$. For $R \gg \kappa^{-1/2}$, the light quark feels the linear part of V_{eam} to a significant extent, and the effects of the Klein paradox become important. We can estimate the probability for pair creation using the Schwinger effective action for a space-time-independent electric field; we find that it is small when R satisfies Eq.(17b). In practice, one can simply check that the wavefunction is well behaved to be sure that these effects are small (in particular, the wavefunction components oscillate wildly for large R). It is interesting to note that in the context of this model we cannot decide whether confinement is a scalar or vector phenomenon. Although confinement in the EAM is, in a sense, primarily due to the vector potential [which guarantees the stability of the bag surface, according to Eq.(17b)], the baglike boundary condition transforms as a *scalar*, and we cannot obtain a solution to the EAM without it.

[31] Note that as long as the "reference" mass in Eq.(22) is not too small, its exact value is not very important for the comparison between $\widetilde{f}_{\text{eam}}(\widetilde{M})$ and $\widetilde{f}_{\text{coul}}(\widetilde{M})\,|_{\xi=0}$, since these functions turn out to be very similar for all values of $\widetilde{M}$ where $\chi_{\text{coul}}(1/\widetilde{M})\,|_{\xi=0}$ can be trusted.

[32] Note that we extend our calculation of $\widetilde{f}_{\text{eam}}(M_P)$ down as far as $M_P \approx \Lambda_{\overline{\text{MS}}}$. Although we expect the light quark wavefunction in the EAM to be reliable at essentially all length scales (unlike the wavefunction in the leading-log potential, which cannot be used near the potential scale mass Λ_{pot}), the application of $\widetilde{f}_{\text{eam}}(M_P)$ to meson masses $\approx \Lambda_{\overline{\text{MS}}}$ must be made with care. This is because our treatment of finite M_Q effects, such as the recoil of the heavy quark, becomes unreliable for very small M_Q. One measure of the reliability of using $\widetilde{f}_{\text{eam}}(M_P)$ at a given mass M_P is the probability of finding the light quark inside the Compton wavelength of the heavy quark $1/M_Q$ [which we have approximated as $1/M_P$; see comments above Eq.(19)]. If the light quark spends an appreciable amount of time "inside" the Compton wavelength of the Q quark, we expect that finite M_Q effects on the wavefunction cannot be neglected. However, even inside a distance as large as $r \approx (2\Lambda_{\overline{\text{MS}}})^{-1}$, the *total* probability of finding the light quark is only about 20%. Thus, $\widetilde{f}_{\text{eam}}(M_P)$ may still give a good approximation to the actual decay constant for meson masses as low as $M_P \approx 2\Lambda_{\overline{\text{MS}}}$.

[33] The function $\widetilde{f}_{\rm eam}(M_P)$ gives a lower bound for $M_{\rm peak}$, and for ratios such as f_B/f_D, for two reasons. (*i*) We have evaluated $\widetilde{f}_{\rm eam}$ by computing the $Q\overline{q}$ overlap at $r_P \equiv 1/M_P$, as an approximation to $r_Q \equiv 1/M_Q$. Since M_P should be greater than M_Q, this means that the overlap used in $\widetilde{f}_{\rm eam}(M_P)$ is at a *smaller* distance than should be used in the true decay constant $f_{\rm rel}(M_Q)$, which has the effect of *increasing* the decay constant. This increase should be more pronounced at smaller meson masses, where the difference between M_P and M_Q should be greatest. For any reasonable connection between M_P and M_Q, the approximation $r_Q \approx r_P$ therefore has the effect of "steepening" the curve for $\widetilde{f}_{\rm eam}$, compared to true decay constant $f_{\rm rel}$, and this should make the meson mass at which $\widetilde{f}_{\rm eam}$ peaks *smaller* than the meson mass at which $f_{\rm rel}$ is maximum. To give a simple example, suppose that $M_P = \lambda M_Q$ where $\lambda \equiv$ constant > 1. Since χ has only one scale (e.g. $\Lambda_{\overline{\rm MS}}$), the peak in $\chi(1/M_P)/\sqrt{M_P}$ occurs at $\lambda^{-1}M_{\rm peak}$, where $M_{\rm peak}$ is the meson mass at which $\chi(1/M_Q)/\sqrt{M_P}$ is maximum. The same reasoning applies to ratios like f_B/f_D; we make a quantitative estimate of this effect for this ratio below Eq.(25). (*ii*) Our definition of $\widetilde{f}_{\rm eam}$ assumed $r_Q \equiv 1/M_Q$, as described in the preceeding argument. This is actually a *lower* bound for r_Q, the distance within which the heavy quark can be localized, and we can repeat the same argument as in (*i*) above to show that the value of $M_{\rm peak}$ and of the ratios f_B/f_D etc. are further increased by using $r_Q > 1/M_Q$.

[34] Note that the development of a peak in $f_{\rm rel}$ is evident at even larger meson masses. For example, our results for $\widetilde{f}_{\rm eam}$ show that $f_{\rm rel}(M_Q)$ changes concavity at a meson mass of at least $9.2\Lambda_{\overline{\rm MS}}$ [=2.1 GeV for $\Lambda_{\overline{\rm MS}}$ =220 MeV].

[35] See the discussion below Eq.(6) and in the beginning of Appendix A.

FIGURE CAPTIONS

Figure 1. Wavefunction components in the effective action model. χ and ϕ are the upper and lower components [see Eq.(3)]. The wavefunction satisfies Eq.(16) [which is coupled non-linearly to Eqs.(15)], subject to the boundary conditions Eqs.(17). The bag radius which satisfies Eq.(17b) is $R \sim 1.16\Lambda_{\overline{\rm MS}}^{-1}$, and is indicated by a vertical line in the figure.

Figure 2. Approximate decay constants in the effective action model [$\widetilde{f}_{\rm eam}(\widetilde{M})$], the leading-log Coulomb potential [$\widetilde{f}_{\rm coul}(\widetilde{M})|_{\xi=0}$] and in the "naive" non-relativistic quark model [$(\widetilde{M}_{\rm ref}/\widetilde{M})^{1/2}$]. $\widetilde{M} \equiv M_P/\Lambda_{\overline{\rm MS}}$, where M_P is the meson mass. $\widetilde{f}_{\rm eam}$ and $\widetilde{f}_{\rm coul}$ are defined in Eqs.(19) and (20). The three functions are normalized to 1 at the "reference mass" $\widetilde{M}_{\rm ref} \equiv 3.5$ [see Eq.(22)],[31] below which the leading-log Coulomb wavefunction cannot be trusted.[32]

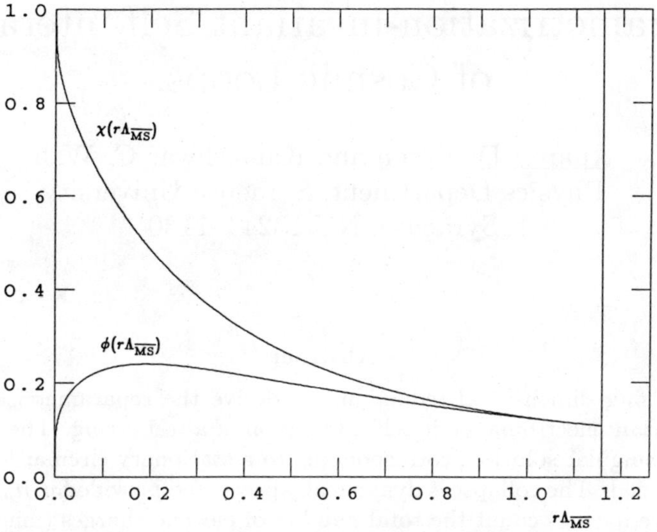

Figure 1.

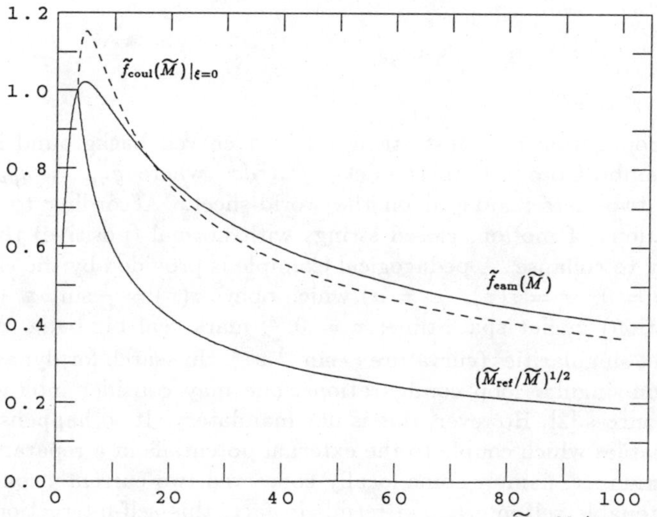

Figure 2.

Reparametrization-invariant Self-interactions of Cosmic Loops.

Aharon Davidson and Kameshwar C. Wali
Physics Department, Syracuse University
Syracuse, NY 13244–1130

Abstract

Using dimensional reduction, we derive the reparametrization-invariant electromagnetic self-interaction of a test string. The exact non-singular solution, corresponding to a stationary circular loop is specified. The collapse is dynamically prevented provided $n_e n_m \neq 0$, where $n_e(n_m)$ count the total number of electric charges (magnetic fluxons). The selection rule $n_e + n_m = 2N$ (N-integer) suggests that a fundamentally-charged loop must carry current and hence acquire spin.

The propagation of a test string [1] in a curved background is governed by the Nambu-Goto action $\int(-\det \tilde{g})^{1/2} d\tau d\sigma$, where $\tilde{g}_{\alpha\beta} = g_{\mu\nu} x^{\mu}{}_{,\alpha} x^{\nu}{}_{,\beta}$ denotes the two-metric induced on the world-sheet. According to the associated equations of motion, closed strings with normal (positive) tension show a tendency to collapse. A pedagogical example is provided by the circular loop ($t = \tau,\ \ z = 0,\ \ r = r(\tau),\ \ \varphi = \sigma$) which obeys $r(\tau) = \frac{1}{\omega}\sin\omega\tau$ (ω-constant of integration) in flat space-time; $\tau = 0, \frac{\pi}{\omega}$ mark real big-bang (crunch)-like world-sheet singularities (curvature $\sim \sin^{-4}\omega\tau$). In search for dynamically stationary non-singular loop configurations, one may consider interactions with external sources.[2] However, this is not mandatory. It so happens that those string variables which couple to the external potentials in a reparametrization-invariant manner, namely some locally conserved two-currents, conspire to establish a tenable self-interaction term. In fact, this self-interaction turns out to have a non-trivial effect even in a flat background. This paper is specifically devoted to the electromagnetic case. The exact solution describing a breathing self-interacting charged superconducting loop [3] is derived and analyzed.

An elegant way to incorporate the electromagnetic interaction without losing reparametrization-invariance is to start from a free five-dimensional [4] Nambu-Goto Lagrangian. While adopting such a unified geometric approach, we choose not to tackle the underlying question whether the fifth coordinate is a reality, or merely a sophisticated technical device. We thus suppress the scalar Kaluza-Klein degree of freedom, focusing on a pure Einstein-Maxwell background. Throughout our analysis we use natural units; in particular, normalized x^5 is identified with a period of 2π.

Let us first briefly explain how the free five-dimensional equations of motion get interpreted by a four-dimensional observer. A key role is played by the dictionary

$$\widetilde{G}_{\alpha\beta} = \tilde{g}_{\alpha\beta} + \Delta_\alpha \Delta_\beta, \tag{1}$$

where$\Delta_\alpha \equiv x^5_{,\alpha} + A_\mu x^\mu_{,\alpha}$, translating the actual world-sheet metric $\widetilde{G}_{\alpha\beta}$ into its effective physical projection $\tilde{g}_{\alpha\beta}$. A_μ represents external potentials. As long as the five-geometry is x^5-independent,

$$Q^\alpha = \frac{\delta}{\delta x^5_{,\alpha}} \sqrt{-\det \widetilde{G}} \tag{2}$$

serves as a conserved ($Q^\alpha_{,\alpha} = 0$) two-current. Eq. (2) can be solved for Δ_α, namely $\Delta_\alpha = \frac{1}{\sqrt{-\det \tilde{g}}} \tilde{g}_{\alpha\beta} \frac{Q^\beta}{M_{in}(Q^2)}$, where

$$M^2_{in} = 1 + \frac{1}{\det \tilde{g}} \tilde{g}_{\alpha\beta} Q^\alpha Q^\beta \tag{3}$$

will play the role of a generalized "intertial mass" [5]. $\Delta_\alpha(Q^\beta)$ is subsequently substituted into the left over equations of motion. These four equations of motion are finally rearranged in the covariant form

$$\left[\frac{\delta\sqrt{-\det \tilde{g}}}{\delta x^\mu_{,\alpha}}\right]_{,\alpha} - \frac{\delta\sqrt{-\det \tilde{g}}}{\delta x^\mu} = \frac{1}{M_{in}(Q^2)} F_{\mu\nu} x^\nu_{,\alpha} Q^\alpha + f^{self}_\mu(Q^\alpha). \tag{4}$$

This formal step, namely isolating the variation of $\sqrt{-\det \tilde{g}}$ on the l.h.s., is the essence of the proper four-dimensional interpretation. The first term on the r.h.s. is recognized as a generalized Lorentz force. Notice the interesting feature that the generalized "inertial mass" $M_{in}(Q^2)$ is not necessarily a constant of motion! The second term on the r.h.s. has no point-particle analog and deserves further attention. However, since we are going to solve the equations of motion in their original form, there is no point in specifying here the rather cumbersome expression [6] for $f^{self}_\mu(Q^\alpha)$. One needs only to be sure of its survival at the flat ($g_{\mu\nu} \to \eta_{\mu\nu}$) limit, and appreciate the facts that f^{self}_μ is (i) $F_{\mu\nu}$-independent, and (ii) even under $Q^\alpha \to -Q^\alpha$.

Consider the circular loop ($t = \tau, \quad z = 0, \quad r = r(\tau), \quad \varphi = \sigma$). The new physics is contained within $x^5(\tau, \sigma)$. Searching for the simplest ansatz of sufficient complexity, we first note that $x^5(\tau,\sigma)$ must be a function of both τ and σ. This is because the trivial cases $x^5(\tau)$ and $x^5(\sigma)$ turn out to suffer from the undesirable radial collapse. Second, we allow for a non-trivial interplay between the external and internal spaces. To be more specific, we do not exclude the topological complication of wrapping the fifth dimension n times while going around the string $(0 < \sigma \leq 2\pi)$. These two elements are present in

$$x^5(\tau,\sigma) = n\sigma + f(\tau) \qquad \text{(n-integer)}, \tag{5}$$

which also makes $\widetilde{G}_{\alpha\beta}$ σ-independent.

The five coupled non-linear equations of motion are satisfied provided $\frac{1}{\omega^2}\dot{r}^2 + \left[\frac{\Omega^2 n^2}{r^2} + r^2\right] = \frac{1}{\omega^2} - \Omega^2 - n^2$,

$$r^4 \dot{f}^2 = \Omega^2\omega^2(r^2+n^2)^2, \tag{6}$$

ω, Ω being constants of integration. The first equation reminds us of a non-relativistic radial harmonic oscillator carrying "angular momentum" $\sim \Omega n$, and "total energy" $\sim \frac{1}{\omega^2} - \Omega^2 - n^2 \geq 2 \mid \Omega n \mid$. It is the $\frac{\Omega^2 n^2}{r^2}$ piece of the effective potential which technically prevents the collapse. This point enters the formal stationary solution

$$r^2(\tau) = a^2 + b^2 \cos 2\omega\tau \tag{7}$$

with the constraint

$$a^4 - b^4 = \Omega^2 n^2, \tag{8}$$

where $a^2 = \frac{1}{2}\left(\frac{1}{\omega^2} - \Omega^2 - n^2\right)$. The self-consistency of this solution requires $\frac{1}{\omega} \geq \mid \Omega \mid + \mid n \mid$, constituting an upper bound on ω. The formula

$$f(\tau) = \pm\Omega\left[\omega\tau + \left|\frac{n}{\Omega}\right| \arctan\left(\frac{|\Omega n|}{a^2+b^2}\tan\omega\tau\right)\right] \tag{9}$$

completes the basic calculation.

Three brief remarks are in order:

(i) Noticing that $-1 + \dot{r}^2 = -\frac{\omega^2}{r^2}(r^2+\Omega^2)(r^2+n^2) < 0$, one may conclude that causality is not violated.

(ii) The special case $\frac{1}{\omega} = |\Omega| + |n|$ corresponds to a static loop configuration, with $r^2 = |n\Omega|$, and hence $\dot{f} = 1$.

(iii) Had we started from a tachyonic world-sheet ($\det \widetilde{G} > 0$), signaling a negative string tension, we would have faced the inflationary solution $r^2(\tau) = a^2 ch 2\omega\tau - b^2$. This solution has the drawback of eventually violating causality.

To gain a deeper insight into the physical meaning of the solution derived, it is essential to address the constants of motion. In particular, we are interested in

$$P_\varphi = n\Omega, \; P_5 = \Omega \tag{10}$$

where $P_M \equiv \frac{1}{2\pi}\int_0^{2\pi} \frac{\delta}{\delta \dot{x}^M}\sqrt{-\det \widetilde{G}}d\sigma$. According to the conventional Kaluza-Klein wisdom, we know that Ω is proportional to the total electric charge of the system. At the local level, such as identification is manifest by

$$Q^\tau = \Omega \quad , \quad Q^\sigma = n\omega\left(1 + \frac{\Omega^2}{r^2}\right). \tag{11}$$

The corresponding four-current $J^\mu = \frac{1}{\sqrt{-\det \tilde{g}}}Q^\alpha x^\mu_{\;,\alpha}$ is given by

$$J^\mu = \frac{1}{\sqrt{1-\dot{r}^2}}(Q^\tau, 0, Q^\tau \dot{r}, Q^\sigma). \tag{12}$$

More exciting is the story told by the non-vanishing and already quantized P_φ. The fact that P_φ is nothing but $\frac{1}{2\pi}\int_0^{2\pi}\left(xP_y^\tau - yP_x^\tau\right)d\sigma$ makes it proportional to the total angular momentum L_z of the system. And since classically $\vec{L}_{e\cdot m\cdot} \sim \vec{r} \times \left(\vec{E} \times \vec{B}\right)$, the presence of an intrinsic magnetic field is essential. With $\Omega \sim$ electric charge, the magnetic effect on P_φ must be concentrated in n. Indeed, the possibility that P_μ^τ contains an electromagnetic piece ΩA_μ^{self}, generated by J_μ, is confirmed by the relation $P_\varphi^\tau = P_5^\tau n$. Altogether, the charged loop superconducts, with

$$\frac{1}{2\pi}\oint A_\mu^{self} dx^\mu = \frac{1}{2\pi}\int_0^{2\pi} A_\varphi^{self} d\varphi = n \tag{13}$$

specifying the total number of magnetic fluxons involved.

The emerging time-periodicity $\Delta\tau = \frac{\pi}{\omega}$ of the world-sheet metric gives rise to the following question: Can a four-dimensional observer actually tell the $\tau = 0$ configuration from the $\tau = \frac{\pi}{\omega}$ one? The naive answer is in the negative, as $r(\tau)$ itself is periodic. But there is still the electromagnetic self-interaction to take into account, and furthermore, one should remember that the fifth-dimensional scale $\sim \frac{h}{e}$ indirectly introduces the Planck constant. Thus, an absolute negative answer to the above question requires

$$\Delta x^5 = 2\pi N \qquad (N - \text{integer}). \tag{14}$$

If Eq. (14) is not satisfied, in the quantum-mechanical language, a non-trivial phase $e^{i\Delta f}$ is picked up. Amazingly, $f(\tau)$ is such that

$$\Delta f = \pm(|\Omega| + |n|)\pi, \tag{15}$$

with the electric and magnetic contributions clearly separated. In turn, the electric charge quantization becomes evident by means of

$$|\Omega| + |n| = 2N. \tag{16}$$

This selection rule is to be automatically satisfied when M_{in} is required to stay a constant of motion.

Although our reasoning is quite novel, the fact that the electric charge Ω takes its values within the sequence $(.. -1, 0, 1, ..)$ is not surprising. It should be emphasized, however, that had we discussed solely charged ($n = 0$) loop (this conclusion may also apply for a charged Kaluza-Klein test point particle), we could have wrongly deduced that one unit of charge corresponds to $|\Omega| = 2$. But Eq. (16) tells us that an odd (even) Ω must be accompanied by an odd (even) n. This in turn gives rise to the following conclusion: The fundamentally-charged ($|\Omega| = 1$) loop necessarily superconducts ($n \neq 0$) and hence acquires spin ($n\Omega \neq 0$).

The generalized "inertial mass" is explicitly given by

$$M_{in}^2 = 1 + \frac{Q^{\tau^2}}{r^2} - \frac{Q^{\sigma^2}}{(1 - \dot{r}^2)} = \frac{r^2 + \Omega^2}{r^2 + n^2}. \tag{17}$$

Notice how the electric (magnetic) contribution tends to increase (decrease) M_{in}. The possibility that M_{in} should be a constant of motion is extremely attractive, but at the moment we lack a compelling reason to support it. The constancy of M_{in} characterizing a static ring ($r = |n\Omega|$) is regarded as a kinematical coincidence. However, the provocative universality of M_{in}, exclusively achieved for $\Omega^2 = n^2$, seems to have a deeper physical origin. As a matter of fact, it follows from a simple covariant constraint

$$\tilde{g}_{\alpha\beta} Q^\alpha Q^\beta = 0, \tag{18}$$

and may be regarded as a generalized equivalence principle. Interestingly, such a principle is in accord with the electric charge-magnetic flux selection rule ($|\Omega| = |n| \Rightarrow |\Omega| + |n| = 2|n|$), allowing the (say) elementary loop to come in two angular momentum states ($n = \pm 1$).

Finally, we would like to make a brief remark concerning the non-Abelian generalization of the ideas exposed here. The full $SU(2)$ analysis has been carried out in detail. [6] Dynamically non-contractible loop configurations both in real space as well as on the intrinsic S_2 manifold exist. The exact solution, in terms of the Weierstrass functions, is quite similar to the Abelian case. An exciting feature is that the up/down iso-configurations are defined by means of a built-in double-well effective potential. Following this line, colored superconductors are conceptually possible.

References

[1] Vilenkin, A., Phys. Rep. 121, 263 (1985).

[2] Balachandran, A.P., Stern, A. and Skagerstam, B.S., Phys. Rev. D20, 439 (1979); Vilenkin, A. and Vachaspati, T., Phys. Rev. Lett. 58, 1041 (1987); Copeland, E., Hindmarsh, M. and Turok, N., Phys. Rev. Lett. 58, 1910 (1987); Spergel, D.N., Piran, T. and Goodman, J., Nucl. Phys. B291, 847 (1987).

[3] Witten, E., Nucl. Phys. B249, 557 (1985); Ostriker, J.P., Thompson, C. and Witten, E., Phys. Lett. B180, 231 (1986).

[4] Nielsen, N.K., Nucl. Phys. B167, 249 (1980); Nielsen, N.K. and Olesen, P., Nucl. Phys. B291, 829 (1987); ibid. Nucl. Phys. B296, 776 (1988).

[5] For the point-particle analog, $M_{in}^2 = 1 - Q^2$, see Davidson, A. and Owen, D., Phys. Lett. B177, 77 (1986).

[6] Davidson, A. and Wali, K.C., SU-4228-385.

Coadjoint Orbits of $Diff(S^1)$, Covariant Operators and the KdV Equation

Wolfgang Scherer
Department of Physics and Astronomy
University of Rochester
Rochester, NY 14627

Abstract

An infinite sequence of higher order differential operators invariant under $Diff(S^1)$ is constructed. These operators are higher order analogues of Hill's operator and they map conformal fields of negative integer and half integer weight to their dual spaces. Some properties and possible applications are discussed. The Korteweg–de Vries equation is formulated as a coadjoint orbit of $Diff(S^1)$.

1 Introduction

In this talk we present some results concerning the diffeomorphism group of the circle $Diff(S^1)$ and its coadjoint representations [1,2]. We present an algorithm to construct an infinite sequence of covariant differential operators for conformal fields. Some properties and possible applications of these operators are discussed. The operators are, in some sense, higher order analogues of Hill's operator which is a covariant operator for conformal fields of weight $-\frac{1}{2}$. Moreover, Hill's operator is used to formulate the Korteweg–de Vries (KdV) equation [3] as a coadjoint orbit of $Diff(S^1)$.

In the first section the coadjoint representation of $Diff(S^1)$ is reviewed and the problem is formulated.

Some known properties of Hill's operator are discussed in section 3 and its higher order analogues are constructed and investigated in section 4. The fifth section contains a new way of writing the KdV equation as a coadjoint orbit.

The results presented here are a report on work that is still in progress, hence, section 4 and 5 contain several questions that still have to be answered and are under investigation.

2 Coadjoint Orbits of $Diff(S^1)$

The group of diffeomorphisms of the circle S^1 is doubly connected and we shall only deal with the orientation preserving part $Diff_+(S^1)$, which we will henceforth call $G := Diff_+(S^1)$.

The Lie–algebra of this group, denoted here by $\mathcal{G}$, is the linear space of vector fields on the circle: $\mathcal{G} = Vect(S^1)$. Its elements $V \in \mathcal{G}$ are represented at a point on S^1 with coordinate θ, $0 \leq \theta \leq 2\pi$ by

$$V(\theta) = v(\theta)\frac{d}{d\theta} =: v(\theta)\partial$$

and they satisfy the algebra

$$[v(\theta)\partial, w(\theta)\partial] = (\, v(\theta)\, w'(\theta) - v'(\theta)\, w(\theta)\,)\, \partial \tag{1}$$

where the prime denotes differentiation with respect to θ. But $\mathcal{G}$ as it stands has no unitary representations and it is therefore its universal extension [4] $\hat{\mathcal{G}} = Vect(S^1) \oplus \mathbf{R}$ that is so important in conformal field theory [5], string theory [6,7] and critical phenomena [8]. $\hat{\mathcal{G}}$ is known as the Virasoro algebra and if we represent $v(\theta)\partial - \imath\alpha c \in \hat{\mathcal{G}}$ (where c is the generator of the center) by the pair (v, α) the algebra may be written as [7]

$$[(v,\alpha),(w,\beta)] = \left(vw' - v'w, \frac{1}{48\pi}\int_{S^1}(vw''' - v'''w)\right). \tag{2}$$

Actually, the most general extension includes the trivial cocycle

$$c_0((v,\alpha),(w,\beta)) = A\int_{S^1}(vw' - v'w) \tag{3}$$

with arbitrary $A \in \mathbf{R}$.

Before we consider the dual of the Virasoro algebra we introduce some notation that will be useful for latter. We define $\Omega_p(S^1)$ as the space of densities of weight p on S^1 or, equivalently, as the space of conformal fields of weight p. $\Omega_p(S^1)$ is also often (formally) described as the space of p-forms: $\Omega_p(S^1) = \{\chi(\theta)(d\theta)^p\}$. Each Ω_p forms a representation of G with $\varphi^{-1} \in G$ being represented by

$$\begin{aligned}\chi(\theta)\,(d\theta)^p \;\overset{\varphi^{-1}}{\longmapsto}\; & \left(\mathcal{R}^{(p)}_{\varphi^{-1}}\chi\right)(\theta)\,(d\theta)^p \\ & = (\varphi'(\theta))^p\;\chi\circ\varphi(\theta)\,(d\theta)^p.\end{aligned} \tag{4}$$

In passing we remark that in this notation $Vect(S^1) = \Omega_{-1}(S^1)$.

Now we return to the consideration of the Virasoro algebra $\mathcal{G}$. Let $\hat{\mathcal{G}}^*$ denote the smooth dual of the Virasoro algebra. Then [9,10]

$$\hat{\mathcal{G}}^* = \Omega_2(S^1) \oplus \mathbf{R} \tag{5}$$

and if we write $q(\theta)\,(d\theta)^2 + \imath bc^* \in \hat{\mathcal{G}}^*$ (where c^* is the dual of c) as a pair (q, b), the pairing between $(v,\alpha) \in \hat{\mathcal{G}}$ and $(q,b) \in \hat{\mathcal{G}}^*$ is given by [7]:

$$\langle (v,\alpha) \mid (q,b)\rangle = \int_{S^1} qv + \alpha b. \tag{6}$$

The coadjoint action of $\varphi^{-1} \in G$ on $(q,b) \in \hat{\mathcal{G}}^*$ was first given by Kirillov [9,11] and has the form:

$$Ad^*_{\varphi^{-1}}(q,b) := \left((\varphi')^2 q \circ \varphi + b\sigma_\varphi, b\right) \tag{7}$$

where σ_φ is the Schwarzian of φ:

$$\sigma_\varphi(\theta) = \frac{1}{2}\frac{\varphi'''(\theta)}{\varphi'(\theta)} - \frac{3}{4}\left(\frac{\varphi''(\theta)}{\varphi'(\theta)}\right)^2 . \tag{8}$$

It has the property

$$\sigma_{\varphi\circ\varrho}(\theta) = (\varrho'(\theta))^2 \sigma_\varphi \circ \varrho(\theta) + \sigma_\varrho(\theta) \tag{9}$$

which assures that (7) is an action.

Actually, the transformation property of (q,b) in (7) is reminiscent of the transformation of gauge fields A under gauge transformations g:

$$A \longmapsto gAg^{-1} + gdg^{-1}. \tag{10}$$

In (10) as in (7) the first term represents the proper tensorial transformation law, whereas the second terms are affine actions in both cases.

In gauge theories (10) can be used to construct covariant derivatives, i.e., a map

$$\mathcal{D} : V_1 \longrightarrow V_2$$

between two vector spaces that carry representations $\mathcal{R}_1$ and $\mathcal{R}_2$, respectively, such that for all g in the gauge group

$$\mathcal{D} \circ \mathcal{R}_1(g) = \mathcal{R}_2(g) \circ \mathcal{D}. \tag{11}$$

From the similarities in (7) and (10), the question arises if it is possible to define covariant derivatives with the help of (q,b).

Indeed, one such covariant derivative is Hill's operator [10,12] and before we construct its generalizations let us review its properties in a manner that is suitable for generalization.

3 Hill's Operator

For brevity, we will restrict our discussion to elements of $\hat{\mathcal{G}}^*$ of the form $\hat{q} := (q, 1)$ from which we can recover general elements (q, b) via $(q, b) = b\widehat{(q/b)}$. For these elements the coefficient of the affine piece in (7) is always one, so we write

$$\left(Ad^*_{\varphi^{-1}}\hat{q}\right)(\theta) = (\varphi'(\theta))^2\, q \circ \varphi(\theta) + \sigma_\varphi(\theta). \tag{12}$$

The operator

$$\mathcal{D}^{(2)}_{\hat{q}} := \partial^2 + q(\theta) \tag{13}$$

is called Hill's operator [13] and has the property [10,12]

$$\begin{aligned} \mathcal{D}^{(2)}_{\hat{q}} : \Omega_{-\frac{1}{2}} &\longrightarrow \Omega_{\frac{3}{2}} \\ \psi &\longmapsto \mathcal{D}^{(2)}_{\hat{q}}\psi \end{aligned} \tag{14}$$

provided $\hat{q}$ transforms as in (12). To see that (14) is true, let $\psi \in \Omega_{-\frac{1}{2}}$, then under $\varphi^{-1} \in G$:

$$\begin{aligned} \left(\mathcal{D}^{(2)}_{\hat{q}}\psi\right)(\theta) \overset{\varphi^{-1}}{\longmapsto} & \left(\mathcal{D}^{(2)}_{Ad^*_{\varphi^{-1}}\hat{q}}\,\mathcal{R}^{(-\frac{1}{2})}_{\varphi^{-1}}\psi\right)(\theta) \\ &= \left[\partial^2 + Ad^*_{\varphi^{-1}}\hat{q}(\theta)\right](\varphi'(\theta))^{-\frac{1}{2}}\,\psi \circ \varphi(\theta) \\ &= (\varphi'(\theta))^{\frac{3}{2}}\left[\left(\frac{1}{\varphi'(\theta)}\partial\right)^2 + q \circ \varphi(\theta)\right]\psi \circ \varphi(\theta) \\ &= (\varphi'(\theta))^{\frac{3}{2}}\left[\frac{d^2}{d\varphi^2} + q \circ \varphi(\theta)\right]\psi \circ \varphi(\theta) \\ &= (\varphi'(\theta))^{\frac{3}{2}}\left[\mathcal{D}^{(2)}_{\hat{q}}\psi\right] \circ \varphi(\theta) \\ &= \left(\mathcal{R}^{(\frac{3}{2})}_{\varphi^{-1}}\mathcal{D}^{(2)}_{\hat{q}}\psi\right)(\theta) \end{aligned} \tag{15}$$

where we used the chain rule

$$\frac{d}{d\varphi} = \frac{1}{\varphi'(\theta)}\frac{d}{d\theta}.$$

The observation (15) indeed proves the claim (14).

A more transparent way of formulating what happened in (15) is to consider the map

$$\begin{aligned} \mathcal{D}^{(2)} : \hat{\mathcal{G}}^* \otimes \Omega_{-\frac{1}{2}} &\longrightarrow \Omega_{\frac{3}{2}} \\ \hat{q} \otimes \psi &\longmapsto \mathcal{D}^{(2)}_{\hat{q}} \psi \end{aligned} \tag{16}$$

Here domain and image carry a representation of G. $\hat{\mathcal{G}}^* \otimes \Omega_{-\frac{1}{2}}$ carries the representation

$$\mathcal{A}^{(-\frac{1}{2})}_{\varphi} := Ad^*_{\varphi} \otimes \mathcal{R}^{(-\frac{1}{2})}_{\varphi} \tag{17}$$

and $\Omega_{\frac{3}{2}}$ carries the representation $\mathcal{R}^{(\frac{3}{2})}_{\varphi}$ as specified in (4). What (15) really tells us is that $\mathcal{D}^{(2)}$ intertwines these representations:

$$\begin{array}{ccc} \hat{\mathcal{G}}^* \otimes \Omega_{-\frac{1}{2}} & \xrightarrow{\mathcal{D}^{(2)}} & \Omega_{\frac{3}{2}} \\ \mathcal{A}^{(-\frac{1}{2})}_{\varphi} \Big\downarrow & & \Big\downarrow \mathcal{R}^{(\frac{3}{2})}_{\varphi} \\ \hat{\mathcal{G}}^* \otimes \Omega_{-\frac{1}{2}} & \xrightarrow{\mathcal{D}^{(2)}} & \Omega_{\frac{3}{2}} \end{array} \tag{18}$$

is commutative, i.e.,

$$\mathcal{D}^{(2)} \circ \mathcal{A}^{(-\frac{1}{2})}_{\varphi} = \mathcal{R}^{(\frac{3}{2})}_{\varphi} \circ \mathcal{D}^{(2)}, \tag{19}$$

and from (11) we conclude that $\mathcal{D}^{(2)}$ is covariant. For general elements $\mathcal{D}^{(2)}_{(q,b)}$ is defined by

$$\mathcal{D}^{(2)}_{(q,b)} = b\,\mathcal{D}^{(2)}_{\widehat{(q/b)}} = b\,\partial^2 + q. \tag{20}$$

Thus, for densities of weight $-\frac{1}{2}$ Hill's operator is the covariant derivative that can be constructed with (q,b).

It is interesting to note that the smooth dual of $\Omega_{-\frac{1}{2}}$ under the G-invariant pairing

$$\langle \psi \mid \alpha \rangle = \int_{S^1} \psi\alpha\,, \quad \psi \in \Omega_{-\frac{1}{2}},\ \alpha \in \Omega^*_{-\frac{1}{2}} \tag{21}$$

can be identified with $\Omega_{\frac{3}{2}}$:

$$\Omega^*_{-\frac{1}{2}} = \Omega_{\frac{3}{2}} \tag{22}$$

such that (16) may be written as:

$$\mathcal{D}^{(2)} : \hat{\mathcal{G}}^* \otimes \Omega_{-\frac{1}{2}} \longmapsto \Omega^*_{-\frac{1}{2}}. \tag{23}$$

Thus, $\mathcal{D}^{(2)}_{(q,b)}$ can be used to define G-invariant bilinear mappings

$$\begin{aligned}(\cdot \mid \cdot)_{(q,b)} : \Omega_{-\frac{1}{2}} \times \Omega_{-\frac{1}{2}} &\longrightarrow \mathbf{R} \\ (\psi_1, \psi_2) &\longmapsto (\psi_1 \mid \psi_2)_{(q,b)}\end{aligned} \tag{24}$$

where

$$(\psi_1 \mid \psi_2)_{(q,b)} := \int_{S^1} \psi_1 \mathcal{D}^{(2)}_{(q,b)} \psi_2. \tag{25}$$

G-invariance, which is a consequence of the property of $\mathcal{D}^{(2)}_{(q,b)}$, means:

$$\left(\mathcal{R}_\varphi^{(-\frac{1}{2})} \psi_1 \mid \mathcal{R}_\varphi^{(-\frac{1}{2})} \psi_2\right)_{Ad^*_\varphi(q,b)} = (\psi_1 \mid \psi_2)_{(q,b)} \tag{26}$$

Although $(\cdot \mid \cdot)_{(q,b)}$ is G-invariant, it is not positive definite as required for an inner product.

This completes the review of some of the properties of Hill's operator in connection with the representations of G. $\mathcal{D}^{(2)}$ is a covariant operator for $\Omega_{-\frac{1}{2}}$ only and it is natural to ask if covariant operators exist for densities of other weight. In the next section this will be answered in the affirmative.

4 Higher Order Analogues

4.1 Construction

We wish to construct differential operators

$$\mathcal{D}^{(k)}_{\hat{q}} = \sum_{j=0}^{k} X^k_j[q](\theta)\, \partial^j \tag{27}$$

where the $X^k_j[q](\theta)$ are polynomials in q and its derivatives (i.e., local functions of q) such that the $\mathcal{D}^{(k)}_{\hat{q}}$ satisfy generalizations of (14). Let us list the desired properties. The analogue of (14) is

$$\mathcal{D}^{(2)}_{\hat{q}} : \Omega_{w_1} \longrightarrow \Omega_{w_2} \tag{28}$$

where w_1 and w_2 are to be determined. With the help of the $\mathcal{D}_{\hat{q}}^{(k)}$ we wish to construct

$$\begin{aligned} \mathcal{D}^{(k)} : \hat{\mathcal{G}}^* \otimes \Omega_{w_1} &\longrightarrow \Omega_{w_2} \\ \hat{q} \otimes \psi &\longmapsto \mathcal{D}_{\hat{q}}^{(k)} \psi \end{aligned} \tag{29}$$

such that $\mathcal{D}^{(k)}$ intertwines the representation

$$\mathcal{A}_{\varphi}^{(w_1)} := Ad_{\varphi}^* \otimes \mathcal{R}_{\varphi}^{(w_1)} \tag{30}$$

and $\mathcal{R}_{\varphi}^{(w_2)}$. In other words, similar to (18) the diagram

$$\begin{array}{ccccc} \hat{\mathcal{G}}^* & \otimes & \Omega_{w_1} & \xrightarrow{\mathcal{D}^{(k)}} & \Omega_{w_2} \\ & \mathcal{A}_{\varphi}^{(w_1)} \Big\downarrow & & & \Big\downarrow \mathcal{R}_{\varphi}^{(w_2)} \\ \hat{\mathcal{G}}^* & \otimes & \Omega_{w_1} & \xrightarrow{\mathcal{D}^{(k)}} & \Omega_{w_2} \end{array} \tag{31}$$

should commute and this can also be written as

$$\mathcal{D}^{(k)} \circ \mathcal{A}_{\varphi}^{(w_1)} = \mathcal{R}_{\varphi}^{(w_2)} \circ \mathcal{D}^{(k)} \tag{32}$$

This is what we call the covariance of the $\mathcal{D}_{\hat{q}}^{(k)}$'s. Let us see if and how (32) can be satisfied. We apply the left hand side for of (32) for a φ^{-1} to $\hat{q} \otimes \psi$:

$$\begin{aligned} \mathcal{D}^{(k)} \circ \mathcal{A}_{\varphi^{-1}}^{w_1} (\hat{q} \otimes \psi)(\theta) &= \left(\mathcal{D}_{Ad_{\varphi^{-1}}^* \hat{q}} \mathcal{R}_{\varphi^{-1}}^{(w_1)} \psi \right) (\theta) \qquad (33) \\ &= \sum_{j=1}^{k} X_j^k \left[Ad_{\varphi^{-1}}^* \hat{q} \right] (\theta) \, \partial^j \left((\varphi'(\theta))^{w_1} \psi \circ \varphi(\theta) \right) \end{aligned}$$

and the right hand side:

$$\begin{aligned} \mathcal{R}_{\varphi^{-1}}^{(w_2)} \circ \mathcal{D}^{(k)} (\hat{q} \otimes \psi)(\theta) &= \left(\mathcal{R}_{\varphi^{-1}}^{(w_2)} \mathcal{D}_{\hat{q}}^{(k)} \psi \right) (\theta) \\ &= (\varphi'(\theta))^{w_2} \left(\mathcal{D}_{\hat{q}}^{(k)} \psi \right) \circ \varphi(\theta) \qquad (34) \\ &= (\varphi'(\theta))^{w_2} \sum_{j=1}^{k} X_j^k [q] \circ \varphi(\theta) \frac{d^j}{d\varphi^j} \psi \circ \varphi(\theta) \\ &= (\varphi'(\theta))^{w_2} \sum_{j=1}^{k} X_j^k [q] \circ \varphi(\theta) \left(\frac{1}{\varphi'(\theta)} \partial \right)^j \psi \circ \varphi(\theta). \end{aligned}$$

Thus, it is easy to see that (32) is equivalent to

$$\sum_{j=1}^{k} X_j^k[Ad^*_{\varphi^{-1}}\hat{q}](\theta)\,\partial^j = (\varphi'(\theta))^{w_2}\left[\sum_{j=1}^{k} X_j^k[q]\circ\varphi(\theta)\left(\frac{1}{\varphi'(\theta)}\partial\right)^j\right](\varphi'(\theta))^{-w_1}. \tag{35}$$

Applying the infinitesimal method to (35) yields

$$w_1 = \frac{1-k}{2} \tag{36}$$

$$w_2 = \frac{1+k}{2} \tag{37}$$

This means that for $\mathcal{D}^{(k)}$ to be covariant, i.e., for $\mathcal{D}^{(k)}$ to satisfy (35) it is necessary that

$$\mathcal{D}^{(k)}_{\hat{q}} : \Omega_{\frac{1-k}{2}} \longrightarrow \Omega_{\frac{1+k}{2}} \tag{38}$$

and surprisingly enough we observe that

$$\Omega^*_{\frac{1-k}{2}} = \Omega_{\frac{1+k}{2}} \tag{39}$$

as was the case for Hill's operator ($k=2$) in (22).

To construct the $\mathcal{D}^{(k)}_{\hat{q}}$ explicitly suppose $\hat{q} \in \hat{\mathcal{G}}^*$ is given then let $\chi \in \Omega_{-\frac{1}{2}}$ be any non-zero solution of Hill's equation:

$$\mathcal{D}^{(2)}_{\hat{q}}\chi = 0 \tag{40}$$

On the interval where χ is different from zero we may define

$$\mathcal{D}^{(k)}_{\hat{q}} := \chi^{-(k+1)}\left(\chi^2\partial\right)^k \chi^{-(k-1)} \tag{41}$$

Clearly χ contains two arbitrary integration constants such that the right hand side of (41) might not be uniquely determined from $\hat{q}$. Using variational techniques, however, it can be shown that the right hand side of (41) is independent of these two constants of integration contained in χ.

To prove that $\mathcal{D}^{(k)}_{\hat{q}}$ satisfies the covariance property (35) we note that as a consequence of (15) Hill's equation for the transformed $\hat{q}$

$$\mathcal{D}_{Ad^*_{\varphi^{-1}}\hat{q}}\bar{\chi} = 0 \tag{42}$$

is solved by

$$\begin{aligned}\bar{\chi}(\theta) &= \left(\mathcal{R}_{\varphi^{-1}}^{(-\frac{1}{2})}\chi\right)(\theta) \\ &= (\varphi'(\theta))^{-\frac{1}{2}}\,\chi\circ\varphi(\theta).\end{aligned} \tag{43}$$

According to the definition (41), the k-th order operator for $Ad^*_{\varphi^{-1}}\hat{q}$ is then

$$\begin{aligned}\mathcal{D}^{(k)}_{Ad^*_{\varphi^{-1}}\hat{q}} &= \bar{\chi}^{-(k+1)}\left(\bar{\chi}^2\partial\right)^k\bar{\chi}^{-(k-1)} \\ &= [(\varphi')^{-\frac{1}{2}}\chi\circ\varphi]^{-(k+1)}\left([(\varphi')^{-\frac{1}{2}}\chi\circ\varphi]^2\partial\right)^k[(\varphi')^{-\frac{1}{2}}\chi\circ\varphi]^{-(k-1)} \\ &= (\varphi')^{\frac{k+1}{2}}\left[(\chi\circ\varphi)^{-(k+1)}\left(\chi\circ\varphi\,\frac{1}{\varphi'}\partial\right)^k(\chi\circ\varphi)^{-(k-1)}\right](\varphi')^{\frac{k-1}{2}}\end{aligned} \tag{44}$$

Equation (44) is nothing but the covariance condition (35) for operators of the form (41). Hence, the operators we constructed are $Diff(S^1)$-covariant.

There is an alternative way to construct the $\mathcal{D}^{(k)}_{\hat{q}}$ using the Miura transformation $q \mapsto v$, which is defined via the Riccati equation

$$q = v' - v^2 \tag{45}$$

for some interval on the circle. On that interval a solution χ of (40) is given by

$$\chi(\theta) = \exp\left(-\int_{\theta_0}^{\theta} v(\lambda)\,d\lambda\right) \tag{46}$$

and for $r \in \mathbf{R}$ we define the operators

$$\Delta_r := \partial + r\,v(\theta) \tag{47}$$

Note that

$$\Delta_r = \chi^r\partial\chi^{-r} \tag{48}$$

These are weight raising operators since

$$\Delta_r : \Omega_{-\frac{r}{2}} \longrightarrow \Omega_{1-\frac{r}{2}} \tag{49}$$

and they have the pseudodifferential inverse

$$\Delta_r^{-1} = \chi^r\partial^{-1}\chi^{-1} \tag{50}$$

Moreover, if $\psi \in \Omega_p$, $\alpha \in \Omega_p^* = \Omega_{1-p}$ and their pairing is similar to the case $p = -\frac{1}{2}$ (21):

$$\langle \psi \mid \alpha \rangle = \int_{S^1} \psi\alpha \tag{51}$$

then,

$$\Delta_r^\dagger : \Omega_{1-\frac{r}{2}}^* = \Omega_{\frac{r}{2}} \longrightarrow \Omega_{-\frac{r}{2}}^* = \Omega_{1+\frac{r}{2}}$$

and

$$\langle \Delta_r \psi \mid \alpha \rangle = \langle \psi \mid \Delta_r^\dagger \alpha \rangle \tag{52}$$

implies

$$\Delta_r^\dagger = -\partial + r\, v(\theta) = -\Delta_{-r} \tag{53}$$

Using the weight raising property (49) and (38) we can construct an operator that maps $\Omega_{\frac{1-k}{2}}$ into $\Omega_{\frac{1+k}{2}}$ by a string of appropriate Δ's:

$$\Delta_{-(k-1)}\, \Delta_{-(k-3)}\, \Delta_{-(k-5)} \cdots \Delta_{k-3}\, \Delta_{k-1} : \Omega_{\frac{1-k}{2}} \to \Omega_{\frac{1+k}{2}} \tag{54}$$

which contains k factors, so k must be an integer.

From (48) we see that indeed

$$\mathcal{D}_{\hat{q}}^{(k)} = \Delta_{-(k-1)}\, \Delta_{-(k-3)}\, \Delta_{-(k-5)} \cdots \Delta_{k-3}\, \Delta_{k-1} \tag{55}$$

In view of (49), we could clearly construct covariant operators that map Ω_s into Ω_p where, except from $p - s \in \mathbf{N}$, p and s are arbitrary by multiplying an appropriate set of Δ's. The resulting operators, however, will not be local in q since they inevitably contain terms linear in v. Only if the product of the Δ's is of the form (55) will there be no linear term in v which is necessary for locality in q. But once $\mathcal{D}_{\hat{q}}^{(k)}$ is of the form (55), it necessarily maps $\Omega_{\frac{1-k}{2}}$ into $\Omega_{\frac{1+k}{2}}$. The previous argument is a different version of the statement that covariance and locality imply (38).

Equations (41) and (55) give two closely related ways of constructing an infinite set of higher order operators $\mathcal{D}_{\hat{q}}^{(k)}$, $k \in \mathbf{N}$ which have the covariance property (35) and satisfy necessary conditions for the X_j^k of (27) to be local in q. However, we have not been able to prove that the construction (41) or (55) is also sufficient for locality, although we strongly believe this to be the case.

The first seven operators are:

$$\begin{aligned}
\mathcal{D}^{(0)}_{\hat{q}} &= 1 \\
\mathcal{D}^{(1)}_{\hat{q}} &= \partial \\
\mathcal{D}^{(2)}_{\hat{q}} &= \partial^2 + q \\
\mathcal{D}^{(3)}_{\hat{q}} &= \partial^3 + 4q\partial + 2q' \\
\mathcal{D}^{(4)}_{\hat{q}} &= \partial^4 + 10q\partial^2 + 10q'\partial + 3q^{(2)} + 9q^2 \\
\mathcal{D}^{(5)}_{\hat{q}} &= \partial^5 + 20q\partial^3 + 30q'\partial^2 \\
&\quad + (18q^{(2)} + 64q^2)\partial + 4q^{(3)} + 6q'q \\
\mathcal{D}^{(6)}_{\hat{q}} &= \partial^6 + 35q\partial^4 + 70q'\partial^3 \\
&\quad + (259q^2 + 63q^{(2)})\partial^2 + (518qq' + 28q^{(3)})\partial \\
&\quad + 225q^3 + 155qq^{(2)} + 130(q')^2 + 5q^{(4)}
\end{aligned} \tag{56}$$

where $q^{(2)} = d^2q/d\theta$ etc. It should be pointed out here that $\mathcal{D}^{(1)}_{\hat{q}}$ and $\mathcal{D}^{(3)}_{\hat{q}}$ are the operators that give the first and second Hamiltonian structure for the KdV equation.

For a general $(q, b) \in \hat{\mathcal{G}}^*$ we define the operators by

$$\mathcal{D}^{(k)}_{(q,b)} := b^n \, \mathcal{D}^{(k)}_{\widehat{q/b}} \tag{57}$$

where n is such that $k = 2n$ or $k = 2n + 1$ depending on whether k is even or odd.

4.2 Properties

Having constructed the sequence of covariant operators, let us list some of their properties that can be derived from their definition.

To begin with, (55) evidently implies the following recursion relation:

$$\mathcal{D}^{(k+2)}_{\hat{q}} = \Delta_{-(k+1)} \mathcal{D}^{(k)}_{\hat{q}} \Delta_{k+1} \tag{58}$$

In order to find the operator adjoint to $\mathcal{D}^{(k)}_{\hat{q}}$ we note that

$$\mathcal{D}^{(k)}_{\hat{q}} : \Omega_{\frac{1-k}{2}} \longrightarrow \Omega_{\frac{1+k}{2}} \tag{59}$$

which implies

$$\mathcal{D}_{\hat{q}}^{(k)\dagger} : \Omega^*_{\frac{1+k}{2}} \longrightarrow \Omega_{\frac{1-k}{2}}$$

Using $\Omega_p = \Omega^*_{1-p}$ this may be written

$$\mathcal{D}_{\hat{q}}^{(k)\dagger} : \Omega_{\frac{1-k}{2}} \longrightarrow \Omega_{\frac{1+k}{2}} \tag{60}$$

and from (53) we obtain

$$\mathcal{D}_{\hat{q}}^{(k)\dagger} = (-1)^k \mathcal{D}_{\hat{q}}^{(k)} \tag{61}$$

such that the $\mathcal{D}_{\hat{q}}^{(k)}$ are (anti-) hermitian for (odd) even k. We may exploit the fact that $\mathcal{D}_{\hat{q}}^{(k)}$ maps $\Omega_{\frac{1-k}{2}}$ into its dual and define a bilinear form on $\Omega_{\frac{1-k}{2}} \times \Omega_{\frac{1-k}{2}}$, which is G-invariant as we did in (25) for the case $k = 2$. Let $\psi_1, \psi_2 \in \Omega_{\frac{1-k}{2}}$; then, we define

$$\begin{aligned} (\cdot \mid \cdot)_{(q,b)} : \Omega_{\frac{1-k}{2}} \times \Omega_{\frac{1-k}{2}} &\longrightarrow \mathbf{R} \\ (\psi_1, \psi_2) &\longmapsto (\psi_1 \mid \psi_2)_{(q,b)} \end{aligned} \tag{62}$$

as in (25). It has the property

$$(\psi_1 \mid \psi_2) = (-1)^k (\psi_2 \mid \psi_1) \tag{63}$$

where (61) has been used. Moreover, this form is G-invariant as stated in (26) which can be seen from the covariance property (35) of the $\mathcal{D}_{\hat{q}}^{(k)}$. Lastly, we note that with the help of (50), the pseudodifferential inverse of $\mathcal{D}_{\hat{q}}^{(k)}$ can easily be constructed:

$$\mathcal{D}_{\hat{q}}^{(k)^{-1}} = \chi^{k-1} (\partial^{-1} \chi^{-2})^k \chi^{k+1} \tag{64}$$

This completes the list of properties directly derivable from the definition. Clearly, in view of possible applications to non–linear integrable systems it would be helpful to have expressions for the commutators of the $\mathcal{D}_{\hat{q}}^{(k)}$ since, if q is time dependent, then as is well known [14,15]

$$\frac{d}{dt} \mathcal{D}_{\hat{q}}^{(2)} = \left[\mathcal{D}_{(q,\frac{8}{3})}^{(3)}, \mathcal{D}_{\hat{q}}^{(2)} \right] \tag{65}$$

leads to the KdV equation for q:

$$\frac{dq}{dt} = \frac{1}{3}q^{(3)} + 2qq' \tag{66}$$

It is not clear if the $\mathcal{D}_{\hat{q}}^{(k)}$ can be used to describe other non–linear integrable systems and work on this is still in progress.

It is also well known [15] that $\mathcal{D}_{\hat{q}}^{(1)}$ and $\mathcal{D}_{\hat{q}}^{(3)}$ are Hamiltonian operators, i.e., if F, H are functions from $\hat{\mathcal{G}}^*$ to the real line the brackets

$$\{F, H\}_1 := \int_{S^1} \frac{\delta F}{\delta q} \mathcal{D}_{\hat{q}}^{(1)} \frac{\delta H}{\delta q} \tag{67}$$

$$\{F, H\}_2 := \int_{S^1} \frac{\delta F}{\delta q} \mathcal{D}_{\hat{q}}^{(3)} \frac{\delta H}{\delta q} \tag{68}$$

are Poisson brackets. Consequently, the question arises whether the $\mathcal{D}_{\hat{q}}^{(k)}$ for odd k have this property, too. This is still being investigated.

At this point it is worthwhile to point out that the Poisson brackets (67) and (68) also yield a representation of the Virasoro algebra as an algebra of functions on $\hat{\mathcal{G}}^*$. To see this, define the Poisson bracket

$$\{F, H\}(q, b) := \int_{S^1} \frac{\delta F}{\delta q} \left(\frac{1}{2} \mathcal{D}_{(q,b/12\pi)}^{(3)} + 2A\, \mathcal{D}_{(q,b)}^{(1)} \right) \frac{\delta H}{\delta q} \tag{69}$$

with arbitrary $A \in \mathbf{R}$ and for each $(v, \alpha) \in \hat{\mathcal{G}}$ we define the function

$$\begin{aligned} P_{(v,\alpha)} : \hat{\mathcal{G}}^* &\longrightarrow \mathbf{R} \\ (q, b) &\longmapsto \langle (v, \alpha) \mid (q, b) \rangle \end{aligned} \tag{70}$$

with the definition (6) and thus, obtain a representation of $\hat{\mathcal{G}}$:

$$\{P_{(v,\alpha)}, P_{(w,\beta)}\}(q, b) = P_{[(v,\alpha),(w,\beta)]}(q, b). \tag{71}$$

The contribution of $\mathcal{D}^{(1)}$ in (69) corresponds to the trivial cocycle (3).

String theory has $Diff(S^1)$ as its invariance group and from geometric quantization one expects the coadjoint orbits to provide information about the quantized version. It is therefore not impossible that the $\mathcal{D}_{\hat{q}}^{(k)}$ can play a role in a geometric formulation of string theory [16].

Finally, as has been shown in [17] for some special cases, the $\mathcal{D}_{\hat{q}}^{(k)}$ are closely related to the Lax operators for higher spin conformal algebras suggesting future work in this direction, too.

5 The KdV Equation as a Coadjoint Orbit

In this section we shall formulate the KdV equation as a coadjoint orbit of $Diff(S^1)$. We begin by reviewing some known facts [18].

Let u, w, ψ be time dependent functions on S^1 : $u = u(t,\theta)$, $w = w(t,\theta)$, $\psi = \psi(t,\theta)$ and let a dot denote differentiation with respect to time and a prime with respect to θ. The system of differential equations for ψ

$$\begin{cases} \mathcal{D}^{(2)}_{\hat{u}}\psi = 0 \\ \dot{\psi} = -\frac{1}{2}u'\psi + u\psi' \end{cases} \tag{72}$$

has solutions ψ if and only if

$$\dot{u} = \frac{1}{2}u''' + 3uu' \tag{73}$$

i.e., if u satisfies the KdV equation [3]. In other words, if for given u solutions ψ exist u must be asolution of the KdV equation.

Now, suppose we wish to solve (73) with initial data

$$u(0,\theta) = q_0(\theta). \tag{74}$$

Let χ_0 be a solution of Hill's equation with potential q_0:

$$\mathcal{D}^{(2)}_{\hat{q_0}}\chi_0 = 0 \tag{75}$$

and let $\varphi_t \in G$ be a time dependent diffeomorphism with

$$\varphi_0(\theta) = \theta \tag{76}$$

The time dependence of φ_t is so far arbitrary. Furthermore, we define

$$q(t,\theta) := Ad^*_{\varphi_t^{-1}}\hat{q}_0(\theta) \tag{77}$$

$$\chi(t,\theta) := \mathcal{R}^{(-\frac{1}{2})}_{\varphi_t^{-1}}\chi_0(\theta) \tag{78}$$

then, one finds as a consequence of (75), (77), (78) and the covariance of $\mathcal{D}^{(2)}_{\hat{q}}$:

$$\begin{cases} \mathcal{D}^{(2)}_{\hat{q}}\chi = 0 \\ \dot{\chi} = \frac{1}{2}\left(\frac{d\varphi_t^{-1}}{dt}\circ\varphi_t\right)'\chi - \left(\frac{d\varphi_t^{-1}}{dt}\circ\varphi_t\right)\chi' \end{cases} \tag{79}$$

In other words, we do know that a solution χ to the system (79) exists. If we choose φ_t such that

$$\left(\frac{d\varphi_t^{-1}}{dt}\circ\varphi_t\right)(\theta) = -q(t,\theta) = -Ad^*_{\varphi_t^{-1}}\hat{q}_0\,(\theta) \tag{80}$$

then (79) becomes

$$\begin{cases} \mathcal{D}^{(2)}_q\chi & = 0 \\ \dot{\chi} & = -\frac{1}{2}q'\chi + q\chi' \end{cases} \tag{81}$$

which is (72) under the replacement $\chi \mapsto \psi$ and $q \mapsto u$; thus, q must be a solution of the KdV equation (73) and from (76) it follows that $q(0,\theta) = q_0(\theta)$.

This shows that q solving the KdV equation follows from φ_t being a solution of (80),i.e.,

$$\begin{cases} \frac{d\varphi_t^{-1}}{dt}\circ\varphi_t & = & -Ad^*_{\varphi_t^{-1}}\hat{q}_0 \\ \varphi_0 & = & id_{S^1} \end{cases} \tag{82}$$

implies that

$$q(t,\theta) = -\frac{d\varphi_t^{-1}}{dt}\circ\varphi_t$$

satisfies

$$\begin{cases} \dot{q} & = & \frac{1}{2}q''' + 3qq' \\ q(0,\theta) & = & q_0(\theta) \end{cases}$$

Equation (82) is thus the KdV equation expressed as a coadjoint orbit of $Diff(S^1)$. Although this formulation does not seem to simplify the search for solutions it may be interesting in the quest for the invariance group of the KdV system. It should perhaps be mentioned here that the above approach resembles very much the formulation of the rigid rotor motion in body–fixed coordinates. In a sense we keep the initial configuration fixed an "move" the circle under it.

Acknowledgements

I would like to thank Prof. S.G. Rajeev for suggesting these investigations and for many instructive discussions. This work was supported by the U.S. Department of Energy Contract No. DE-AC02-76ER13065.

References

[1] Scherer, W., University of Rochester Preprint, UR-1046, to appear in *Lett. Math. Phys.*

[2] Scherer, W., *Canonical Quantization of Constrained Systems and Coadjoint Orbits of* $Diff(S^1)$, Ph.D. Thesis, University of Rochester, 1988.

[3] Korteweg, D.J. and de Vries, G., *Phil. Mag.* **39** (1885) 422.

[4] Gelfand, I.M. and Fuks, D.B., *Funct. Anal. its Appl.* **2** (1968) 92.

[5] Polyakov, A.M., *JETP Lett.* **12** (1970) 381; Belavin, A.A, Polyakov, A.M. and Zamolodchikov A.B., *Nucl. Phys.* **B241** (1984) 333.

[6] Bowick, M. and Rajeev, S., *Phys. Rev. Lett.* **58** (1987) 535; *Nucl. Phys.* **B293** (1987) 348.

[7] Witten, E., *Coadjoint Orbits of the Virasoro Group*, Princeton University Preprint, PUPT-1061.

[8] Friedan, D., Qiu, Z. and Shenker, S., *Phys. Rev. Lett.* **52** (1984) 1575 and in: Vertx Operators in Mathematical Physics, Lepowsky, J. et al. (ed.), Springer, Berlin, 1984.

[9] Kirillov, A.A., *Funct. An. its Appl.* **15** (1981) 135.

[10] Segal, G., *Commun. Math. Phys.* **80** (1981) 301.

[11] Kirillov, A.A. in: Twistor–Geometry and Non–Linear Systems, Doebner, H.D. et al. (ed.), *Lecture Notes in Mathematics* **970**, Springer, Berlin, 1982.

[12] Lazutkin V. and Pankratova, T., *Funct. An. its Appl.* **9** (1975) 306.

[13] Magnus, W. and Winkler, S., Hill's Equation, Interscience, New York, 1966.

[14] Lax, P.D., *Comm. Pure Appl. Math.* **21** (1968) 467.

[15] Abraham, R. and Marsden, J.E., *Foundations of Mechanics*, Benjamin/Cummings, Reading, Massachusetts, Second Edition, 1978.

[16] Rajeev, S., private communication.

[17] Mathieu, P., *Extended Classical conformal algebras and the second Hamiltonian structure of Lax equations*, Université Laval preprint.

[18] Ablowitz, M.J., Kaup, D.J., Newell, A.C. and Segur, H., *Phys. Rev. Lett.* **31** (1973) 125.

A Study of Hermitian Quark Mass Matrices.

J.A. Robinson*†
Physics Department
McGill University
Montreal, P.Q., Canada, H3A 2T8.

T.G. Rizzo‡
Ames Laboratory and Department of Physics
Iowa State University
Ames, Iowa, U.S.A. 50011

Abstract

In this talk review a recent analysis of three generation Hermitian quark mass matrices and the Kobayashi-Maskawa mixing matrices which they produce. It is found that constraints obtained from the CP violation parameter $\epsilon, B^o - \bar{B}^o$ mixing, and the recent observation of charmless B-decays are very restrictive and effectively eliminate the Fritzsch Scheme as a realistic three generation model. On the other hand, we find that several of the other mass matrices examined are in good agreement with the most recent experimental and theoretical limits.

1 INTRODUCTION

The three generation Standard Model[1] (SM) can account for a wide range of phenomena. Indeed, there is not a single piece of experimental data in conflict with its predictions. However, a common criticism of the SM is that it contains an exceedingly large number of unknown free parameters. A minimal counting

*Talk presented at the Tenth Annual Montreal-Rochester-Syracuse-Toronto Meeting, Toronto, Canada, May, 1988

†This work was supported in part by NSERC and in part by the Quebec Department of Education, FCAR.

‡This work was supported by the U.S.Department of Energy, Contract No. W-7405-Eng-82, Office of Energy Research (KA-01-01), Division of High Energy and Nuclear Physics.

yields at least eighteen a priori undetermined parameters. Certainly, it would be preferable to eliminate as many free parameters from the theory as possible. One scenario is to assume that the Kobayashi-Maskawa[2] (KM) mixing matrix, whose elements we will denote by V_{ij}, arises from a particular quark mass matrix. A mass matrix which is non-diagonal in the weak interaction basis, will, upon rotation of the fields, produce a unitary mixing matrix in the charged current sector, which may be identified as the KM matrix. Thus, if the charge $+2/3$ and $-1/3$ quark mass matrices are specified then the charged current mixing matrix is uniquely determined.

A popular ansatz for the quark mass matrix is that proposed by Fritzsch[3]. This scenario has been found to produce phenomenologically acceptable KM elements[4]. It has been pointed out[5], however, that this form of the quark mass matrix has some difficulty accounting for results[6] from $B^o - \bar{B}^o$ mixing. We corroborate this conclusion, and more importantly we will show that the limits placed on the ratio $\mid V_{ub}/V_{cb} \mid$ from the observation of charmless B decays[7] appear to eliminate this scheme as a viable three generation model. In a search for alternatives to the Fritzsch scenario, we have previously completed an analysis of a very large class of Hermitian mass matrices[8,9]. In Ref. 8 we found that several matrices are in good agreement with the present constraints on the magnitudes of the elements of the KM matrix. In Ref. 9, we continued our study of these mass matrices and extended our analysis to include the experimental constraints on the CP violation parameters ϵ and ϵ'/ϵ, the $B^o - \bar{B}^o$ mixing parameter x_d, and the ratio $\mid V_{ub}/V_{cb} \mid$. In this talk we will review our work in Ref. 8 and in Ref. 9. For further details the reader is referred to these works.

2 QUARK MASS MATRICES

In terms of weak eigenstates the quark-mass sector of the SM can be written

$$L_m = \bar{U}^o_L M^o_u U^o_R + \bar{D}^o_L M^o_d D^o_R + \text{h.c.}, \tag{1}$$

where $U(D)$ is a column vector representing all charge $+2/3(-1/3)$ quarks. The complex matrices M^o_u and M^o_d are the $+2/3$ and $-1/3$ non-diagonal mass matrices, respectively. In this basis the charged current is

$$J^\mu_{cc} = \bar{U}^o \gamma^\mu (1-\gamma_5) D^o. \tag{2}$$

In rotating the quark fields from the weak interaction basis to the physical basis (where M_u and M_d are diagonal)

$$L_m = \bar{U}_L M_u U_R + \bar{D}_L M_d D_R + \text{h.c.} \tag{3}$$

a unitary matrix, V_{KM}, is generated in the charged current sector

$$J^\mu_{cc} = \bar{U} \gamma^\mu (1-\gamma_5) V_{KM} D. \tag{4}$$

This process is carried out by first "rotating away" the complex phases in M^o to obtain the real matrix M_R. For the matrices we will consider all of the phases in M^o can be absorbed into the quark fields. (However, one should note that this is not always possible for a general Hermitian matrix). Once the real matrix M_R is obtained it is straightforward to obtain M via an orthogonal transformation O. If $O_u(O_d)$ is the orthogonal matrix which diagonalizes the real form of $M_u^o(M_d^o)$, then, for the matrices considered here, the matrix V_{KM} is given by

$$V_{KM} = O_u^+ \begin{bmatrix} 1 & 0 & 0 \\ 0 & e^{i\sigma} & 0 \\ 0 & 0 & e^{i\tau} \end{bmatrix} O_d . \tag{5}$$

The details of this calculation can be found elsewhere[3,4,8]. The important point to remember is that the elements of V_{KM} are completely determined in terms of the quark masses and the two phases σ and τ. These two phases are related to the unknown phases existing in $(M^o)_{u,d}$ and are in general arbitrary. It may be that within a particular model these phases are fixed by the structure of the theory, such as by the Higgs sector.

In Ref. 8, we classified all Hermitian mass matrices consisting of two or more zero elements and three complex parameters (A, B, and C). We allow only three parameters so that their respective magnitudes can be uniquely determined (at least numerically) in terms of quark masses. These magnitudes are calculated by using the three invariants obtained by expanding the eigenvalue equation $\det[M - \lambda I] = 0$, where the eigenvalues λ_i are the quark masses. For the case of three- and four-zero matrices we found[8] that only a few are capable of producing values of the KM matrix elements which are consistent with direct experimental data. These mass matrices are shown in Table I. By employing the same analysis as we did for the three-zero case, we have recently found that only one type (see Table I) of two-zero matrix produces reasonable values for the KM elements. The matrices shown in Table I are the ones which will be further discussed in this talk. For a detailed account of the procedure used to arrive at these matrices the reader is referred to Ref. 8.

In order to calculate the elements of the KM matrix we need to insert proper values for the quark masses into our expressions for A, B, and C. The appropriate mass to insert into this calculation is the mass parameter which appears in the Lagrangian of the theory, the so called running mass $m_i(\mu)$. In determining V_{KM} we use the running quark masses evaluated at the scale $\mu = 1$ GeV [10]

$$\begin{aligned} m_u &= (5.1 \pm 1.5)\text{MeV} \quad m_c = (1.35 \pm 0.05)\text{GeV} \\ m_d &= (8.9 \pm 2.6)\text{MeV} \quad m_s = (175. \pm 55.0)\text{MeV} \quad m_b = (5.3 \pm 0.1)\text{GeV} \end{aligned} \tag{6}$$

and $\quad \frac{m_d}{m_u} = 1.76 \pm 0.13 \qquad \frac{m_s}{m_d} = 19.6 \pm 1.6$.

These are the quark masses renormalized in the modified-minimal-subtraction scheme with $\Lambda_{\overline{MS}} = 100$ MeV. We leave the t-quark mass as a free parameter.

At an arbitrary scale μ the running mass $m(\mu)$ is given by Ref. 9.

$$m(\mu) = \bar{m}\left[1 - \frac{2\beta_1\gamma_0}{\beta_0^3}\frac{lnL+1}{L} + \frac{8\gamma_1}{\beta_0^2 L}\right]\left[\frac{L}{2}\right]^{\frac{-2\gamma_0}{\beta_0}} \tag{7}$$

where,

$$\beta_0 = 11 - \frac{2}{3}N_f \quad ; \quad \gamma_0 = 2$$
$$\beta_1 = 102 - \frac{38}{3}N_f \quad ; \quad \gamma_1 = \frac{101}{12} - \frac{5}{18}N_f$$
$$L = ln(\mu^2/\Lambda^2_{\overline{MS}}).$$

Here $\bar{m}$ is the renormalization group invariant mass and N_f is the number of quark flavors. In the calculation of physical processes the correct quark mass to use is the running mass evaluated at the scale $\mu = m$, $m(\mu = m)$. So, in our calculations below the correct masses to use are $m_t(m_t)$ and $m_c(m_c)$. Our future notation will be that $m = m(\mu = m)$ and in particular the t-quark mass displayed in our figures is understood to be $m_t(m_t)$. Next we will discuss some of the recent limits which have been placed on the mass of the t-quark.

Even though there has not been a direct observation of the t-quark, one can still place limits on its mass through various indirect methods. From a comprehensive analysis of existing weak neutral current data it is possible to set an upper limit on m_t [11] : $m_t < 200$ (180) GeV, at 90% C.L. assuming that the Higgs mass is less than 1 TeV (100 GeV). A stronger upper limit of $\simeq 65$ GeV can be placed on m_t from data[12] on the ratio of the widths of the W and Z bosons[13,14]. Direct searches at PETRA and TRISTAN yield the lower limits of $m_t > 23$ GeV [15] and $m_t \geq 25.8$ GeV [16], respectively. A model dependent lower limit has been announced by the UA1 Collaboration[17] : $m_t > 45$ GeV. In our analysis we will allow the whole range $m_t = 23 - 200$ GeV.

3 CONSTRAINTS

In this section we discuss the present experimental limits on the quark mixing sector of the three generation SM.

3.1 KM Matrix

The magnitude of the elements of the KM matrix are quite constrained. The present "best" values are[18]

$$\begin{array}{ll} \mid V_{ud} \mid = 0.9748 \pm 0.0010 & \mid V_{cd} \mid = 0.207 \pm 0.024 \\ \mid V_{us} \mid = 0.220 \pm 0.002 & \mid V_{cs} \mid = 0.95 \pm 0.14 \\ \mid V_{ub} \mid < 0.012 & \mid V_{cb} \mid = 0.043^{+0.006}_{-0.008}. \end{array} \tag{8}$$

There have been some very important recent results that provide further constraints on these elements. By determining the ratio $\Gamma(b \to ul\nu)/\Gamma(b \to cl\nu)$ from semileptonic B decays a 90% C.L. upper limit on the ratio $\mid V_{ub}/V_{cb} \mid$ has been obtained[19]

$$\left| \frac{V_{ub}}{V_{cb}} \right| \leq 0.20 \, . \tag{9}$$

Very recently charmless B decays have been observed[7] resulting in the lower bound

$$\left| \frac{V_{ub}}{V_{cb}} \right| \geq 0.07 \, . \tag{10}$$

Thus, Equations (8), (9), and (10) taken together are the most recent constraints on the magnitudes of the KM elements and we will use these in our analysis below. One should note that due to the unitarity of the KM matrix, the other elements which are not directly measured are also constrained.

3.2 CP-violating Parameters ϵ and ϵ'/ϵ

The main contribution to the CP-violation parameter ϵ arises from the box diagram. The result is

$$\mid \epsilon \mid = -\frac{BG_F^2 m_W^2 m_K f_k^2}{12\sqrt{2}\ \pi^2 \Delta M_{L,S}} Im \left[\sum_{i,j=2}^{3} \lambda_i \lambda_j \eta_{ij} E(x_i, x_j) \right] , \tag{11}$$

where $E(x_{ij} x_j)$ is a well known function[20] and the λ_i and x_i are defined by $\lambda_i \equiv V_{is}^\star V_{id}$ and $x_i = (m_i/m_w)^2 \quad i = c, t$. In Eq. (11) the u-quark contribution has been eliminated by using the unitarity of the KM matrix. As input into our calculations we use the value[21]

$$\mid \epsilon \mid = 2.274 \times 10^{-3}. \tag{12}$$

Also, as input into Eq. (11) we use the values $f_K = 160$ MeV and $\Delta M_{L,S} = 3.52 \times 10^{-15}$ GeV [21]. The values we use for QCD correction factors are $\eta_{22} = 0.69$, $\eta_{33} = 0.59$, and $\eta_{23} = 0.41$ [22]. The largest uncertainty in Eq. (11) comes from the parameter B. This parameter is defined in terms of the hadronic matrix element

$$\frac{4}{3}Bf_K^2M_K = \langle K^o \mid [\bar{s}\gamma^\mu(1-\gamma_5)d]^2 \mid K^o\rangle.$$

The vacuum insertion approximation result is $B = 1.0$ [23] whereas, a recent calculation, using QCD sum rules gives; e.g., $B = 0.84 \pm 0.08$ [24]. The calculation of B has a long history of varied results[25], so we choose two different ranges (one loose and one tight) for this parameter

$$0.33 \leq B \leq 1.75 \qquad \text{"loose"} \tag{13a}$$

$$0.33 \leq B \leq 1.0 \qquad \text{"tight."} \tag{13b}$$

We will include in our results both of the ranges Eq. (13a) and Eq. (13b).

Another measure of CP violation is the quantity ϵ'/ϵ. This ratio is experimentally determined to be

$$\frac{\epsilon'}{\epsilon} = \begin{cases} (\ 1.7 \pm 8.4) \times 10^{-3} & \text{(Yale-BNL)}^{26]} \\ (-4.6 \pm 5.8) \times 10^{-3} & \text{(Chicago-Saclay)}^{27]} \\ (\ 3.5 \pm 3.6) \times 10^{-3} & \text{(E731 "Preliminary")}^{28]} \\ (\ 3.5 \pm 1.4) \times 10^{-3} & \text{(NA31 "Preliminary")}^{29]} \end{cases} \tag{14}$$

Unfortunately, there is a great deal of theoretical uncertainty involved in the calculation of this ratio. It is believed that the main contribution to ϵ'/ϵ comes from the "strong penguin" diagram however there are also other contributions to this ratio including "long-distance" effects. Recent studies indicate that these contributions are substantial and cannot be neglected[30,31]. These effects can be included by writing

$$\left|\frac{\epsilon'}{\epsilon}\right| = \left|\frac{\epsilon'}{\epsilon}\right|_{sp}(1 - \Omega_{emp} - \Omega_{\eta\eta'}), \tag{15}$$

where Ω_{emp} and $\Omega_{\eta\eta'}$ represent the QED penguin and $\pi_o-\eta-\eta'$ mixing contributions, respectively. The QED contribution is given by $\Omega_{emp} \simeq -0.002$ [Ref. 30]. Two recent estimates of the long-distance contribution are $\Omega_{\eta\eta'} \simeq 0.27$ [Ref. 30] and $\Omega_{\eta\eta'} = 0.40 \pm 0.06$ [Ref. 31] so that the result for ϵ'/ϵ is reduced by about 30% from the "strong penguin" calculation alone. Because of the large number of theoretical and experimental uncertainties associated with this parameter we do not use it as a constraint in our analysis. We do, however, calculate this ratio and compare it to the experimental values shown in Eq. (14).

3.3 $B^o - \bar{B}^o$ Mixing

By studying the decays of the B mesons produced in the reaction

$$e^+e^- \rightarrow \Upsilon(4s) \rightarrow B^o(\bar{b}d) + \bar{B}^o(b\bar{d})$$

the ARGUS Collaboration has measured substantial $B^o - \bar{B}^o$ mixing[6]. The strength of the mixing is given by the parameter r which is defined by

$$r = \frac{N(B^o B^o) + N(\bar{B}^o \bar{B}^o)}{N(B^o \bar{B}^o)} = \frac{(\Delta M)^2 + (\Delta\Gamma/2)^2}{2\Gamma^2 + (\Delta M)^2 - (\Delta\Gamma/2)^2}.$$

The mass difference between the two mass eigenstates is Δm, and, $\Gamma^{-1} = \tau_B$, where τ_B is the B lifetime. Since $(\Delta\Gamma/2)^2 \ll (\Delta M)^2$, r can be accurately approximated by

$$r = \frac{x_d^2}{x_d^2 + 2} \quad , \quad x_d = \frac{\Delta M}{\Gamma}. \tag{16}$$

The observed value of r is[6]

$$r = 0.21 \pm 0.08 \quad , \quad x_d = 0.73 \pm 0.18 \ . \tag{17}$$

The parameter x_d is given by $x_d = 2 \mid M_{12} \mid \tau_B$ where $\mid M_{12} \mid$ is calculated from the box diagram connecting B^o to $\bar{B}^o$.

$$\mid M_{12} \mid = \frac{B_B f_B^2 G_F^2 m_W^2 m_B}{12\pi^2} \left| \sum_{i,j=2}^{3} \lambda_i \lambda_j \eta_{ij} E(x_i, x_j) \right| , \tag{18}$$

where $E(x_i, x_j)$ is as in Eq. (11) and $\lambda_i = V_{ib}^\star V_{id}$.

The product $f_B^2 B_B$ in this expression is not well determined but is expected to lie in the range $f_B^2 B_B = (0.15 \pm 0.05)^2$, however, recent calculations seem to favor the lower end of this range[32]. Also, τ_B is only known at the level of $\simeq 10\%$ with the most recent world average value being $\tau_B = (1.19 \pm 0.11) \times 10^{-12}$ sec [7]. In order to take these uncertainties into account we define the following quantity

$$R = \frac{x_d}{f_B^2 B_B \arg(\tau_B)}, \tag{19}$$

where $\tau_B = \arg(\tau_B) \times 10^{-12}$ sec. Substituting numerical values into Eq. (19) we obtain the bounds

$$4.11 \leq R \leq 66.04 \qquad \text{“loose”} \tag{20a}$$

$$6.92 \leq R \leq 55.08 \qquad \text{“tight.”} \tag{20b}$$

The range in Eq. (20a) is obtained by allowing the full range given by the (1σ) errors on the quantities x_d, $f_B^2 B_B$, and τ_B. The range in Eq. (20b) is obtained by allowing $f_B^2 B_B$ and τ_B to take on the same values as in Eq. (20a) and allowing x_d to take on values as large as two standard deviations (2σ) away from its central value. One should be aware that Eq. (20a) is a very conservative constraint. It is unlikely that the three quantities in question would conspire in such a way as actually to allow values at the edge of this range. Our results will include both of the ranges in Eq. (20).

4 ANALYSIS

In this section we explain the procedure used to obtain our results.

From Eqs. (1)–(5) we see that if the quark mass matrix is specified then the elements of V_{KM} are uniquely determined. However, by specifying the form of the mass matrix we do not uniquely determine V_{KM}. This is clear from Eq. (5) since a priori σ and τ are free parameters. Also, since there are uncertainties in the values of the quark masses the non-zero elements A, B, and C will also vary. To determine the full range of values for the V_{ij} we let σ and τ run over the range 0 to 2π in steps of 0.01π radians. The uncertainties in the light quark masses are accounted for by letting m_u, m_d, and m_s take on the full range of values allowed by Eq. (6).

In our analysis, we first require that the matrix V_{KM} satisfy the constraints Eqs. (8), (9), and (10). We then further require that the values of B and R calculated via Eqs. (11) and (18) satisfy as a minimum the loose constraints Eqs. (13a) and (20a). If for some set of values of σ, τ, and m_i, a mass matrix can not produce a mixing matrix satisfying these minimal constraints, it is discarded.

5 RESULTS AND DISCUSSION

In the following we will present our results separately for each matrix.

5.1 The Fritzsch Matrix

The structure of the Fritzsch mass matrix (M_F) is shown in Table 1. For values of σ and τ in the ranges $\sigma \simeq \tau \simeq (0.5 \pm 0.1)\pi$ and $\sigma \simeq \tau \simeq (1.5 \pm 0.1)\pi$ the predicted values for the V_{ij} are in good agreement with Eqs. (8) and (9) except that for a very small range of m_t values the limits Eqs. (10), (13), and (20) are not satisfied. Since the range of allowed m_t is so small, we will (for purposes of discussion) only require that Eqs. (8) and (9) be passed by this theory. We see in Figs. 1, 2, and 3 that for $m_t > 80$ GeV even these constraints are not satisfied. Fig. 1 shows that the lower limit of $R > 4.11$ is met for $m_t \simeq 55 - 80$ GeV. As m_t nears 80 GeV the value of R nearly satisfies the tight constraint $R > 6.92$, but, as shown in Fig. 2, as m_t is further increased the ratio $\mid V_{ub}/V_{cb} \mid$ falls below the bound $\mid V_{ub}/V_{cb} \mid > 0.07$. Only for $m_t \simeq 50 - 60$ GeV does M_F yield $\mid V_{ub}/V_{cb} \mid$ values greater than this lower limit. In Fig. 3 we see that in this mass region B is rather large, $B \gtrsim 1.2$. Hence, Figs. 1–3 show that the loose R and B limits can only be satisfied for a very tiny range of m_t values and the tight constraints Eqs. (13b) and (20b) are never satisfied. It is important to note that in order to satisfy the lower limit Eq. (20a) requires all three quantities $f_B^2 B$, τ_B, and x_d to be very near the edge of their allowed ranges. The maximum

R value that M_F predicts for $m_t \simeq 55 - 60$ GeV is $R \simeq 4.5$. Maximizing $x_d(x_d = 1.09)$ and minimizing $\tau_B(\tau_B = 1.08$ psec) implies that $R \simeq 4.5$ only for $f_B^2 B \gtrsim (0.19)^2$. However, as mentioned in Section III recent calculations favor values at the lower end of the range $f_B B^{1/2} = 0.10 - 0.20$. Summarizing, in order for M_F to satisfy the loose constraints the t-quark mass must lie in the range $m_t \simeq 55 - 60$ GeV and all other parameters must take on extreme values: $B \simeq 1.2$, $\tau_B \simeq 1.3$ psec, $r \simeq 0.05$, and $f_B^2 B \simeq (0.19)^2$. (Also note that although not explicitly shown here, M_F also requires m_s to be an extreme minimum value of $m_s \simeq 120.$ MeV.) If the constraint from charmless B-decays is removed the situation is somewhat improved, especially for $m_t \simeq 80$ GeV[33].

5.2 Three-zero Mass Matrices

In our earlier work we found several mass matrices with three zero elements which could satisfy the constraints Eqs. (8) and (9). Of these matrices those passing the new constraints Eqs. (10), (13), and (20) are included in Table I. Our results for these matrices are shown in Table II and in Figs. 4–15. These results are obtained from KM matrices which as a minimum must satisfy Eqs. (8), (9), (10), (13a), and (20a). We first turn our attention to the matrix M_5.

5.2.1 M_5

In Fig. 4 we have plotted the allowed values of R for various m_t values. The two dotted lines correspond to the two lower limits in Eqs. (20a) and (20b). The shaded region represents values of R arising from KM matrices which also yield values of B in the range Eq. (13b). The unshaded region represents values satisfying only Eq. (13a). Therefore, Fig. 4 shows that for $m_t \simeq 55 - 75$ GeV, M_5 predicts R values in good agreement with $B^o - \bar{B}^o$ mixing data. In Fig. 5 we see that this matrix can accommodate nearly the entire range of B values allowed by Eq. (13). In this figure, the shaded region represents B values that have been calculated from KM matrices which produce values of R lying in the range Eq. (20b). In Fig. 6 we plot $| V_{ub}/V_{cb} |$ for various m_t values. Here the shaded region represents values obtained from KM matrices which predict values of R and B satisfying both of the tight constraints, Eqs. (13b) and (20b). The unshaded region represents points which only satisfy Eqs. (13a) and (20a). Fig. 6 shows that if the present experimental upper limit on $| V_{ub}/V_{cb} |$ can be lowered then this matrix could be elininated from further consideration. In Table II we have tabulated representative ranges of ϵ'/ϵ for various values of m_t. Table II represents values of ϵ'/ϵ for which the corresponding B and R values pass the loose constraints. We see that this matrix yields values for ϵ'/ϵ which are in good agreement with Eq. (14).

5.2.2 $M_6(A_2)$

This matrix is distinguished from $M_6(A_3)$ by the value for A used in the mass matrix. In the determination of this parameter from the eigenvalue equation one obtains a cubic equation in A, hence the different roots A_2 and A_3 [Ref. 8]. For $m_t \gtrsim 70$ GeV, $M_6(A_2)$ easily satisfies all of the constraints that we have imposed. This is fairly surprising since we previously found that this matrix has a very small allowed region in $\sigma - \tau$ space. We see in Fig. 7 that all values of R in the range Eq. (20) are allowed, note that smaller m_t values tend to yield smaller R values. As we will see below, this is the only matrix we have considered for which the upper limits in Eq. (20) are even approached. In addition, Fig. 8 shows that a wide range of B values are also allowed. In Fig. 9 we see that this matrix allows the full range of values for the ratio $| V_{ub}/V_{cb} |$ when $m_t < 90$ GeV. For $m_t > 100$ GeV the predicted values for $| V_{ub}/V_{cb} |$ are always $\gtrsim 0.11$. We have closely examined the region at $m_t \simeq 90$ GeV where there is a sharp change in the shape of the displayed region. Unlike the other matrices it appears that this matrix is quite sensitive to very small changes in the values of the input parameters. By doubling the number of steps used in scanning over the quark mass ranges and $\sigma - \tau$ space we obtained a far smoother lower bound over the range $m_t \simeq 90 - 120$ GeV. This implies that this sharp jump is simply an artifact of our resolution. Of course one should expect a slight change in our results by reducing the resolution used for all of the matrices examined, however, it appears that the $| V_{ub}/V_{cb} |$ ratio for this particular matrix is extremely sensitive to this resolution. Note, however, that the regions shown in Figs. 7 and 8 do not show such a sharp change at $m_t \simeq 90$ GeV. The values of ϵ'/ϵ shown in Table II are outside of the 1σ range of the $NA31$ limits, however, one can not rule out this matrix on this basis. Recall that the calculation of this ratio involves a great deal of uncertainty. Also, the predicted values do fall well within the range obtained by allowing a 2σ deviation from the central value: $0.7 \times 10^{-3} \leq \epsilon'/\epsilon \leq 6.3 \times 10^{-3}$.

5.2.3 $M_6(A_3)$

In our previous work we found that this matrix satisfies Eqs. (8) and (9) over a large region of $\sigma - \tau$ parameter space. Thus, one might expect that this matrix would satisfy the new constraints from ϵ, x_d, and $| V_{ub}/V_{cb} |$ quite well. We see in Figs. 10 and 11 that this is indeed true. However, the tight $B^o - \bar{B}^o$ constraints are satisfied only for $m_t \gtrsim 95$ GeV. For values as low as $m_t \simeq 70$ GeV the loose $B^o - \bar{B}^o$ constraints can also be satisfied. We see in Fig. 12 that this matrix only allows a very narrow range of values for $| V_{ub}/V_{cb} |$. In fact, if $| V_{ub}/V_{cb} | \gtrsim 0.074$, this matrix could be eliminated. Another feature of this matrix is that it requires all light quark mass values to be near their allowed upper limits.

5.2.4 $M_9(A_3)$

This matrix does a poor job of satisfying our constraints. Actually the tight constraints on R can be satisfied (as can be seen in Figs. 13 and 14) only for a very tiny region at $m_t \simeq 80$ GeV. In Fig. 15 we see that as m_t gets larger $\mid V_{ub}/V_{cb} \mid$ gets smaller, until at $m_t = 80$ GeV, $\mid V_{ub}/V_{cb} \mid$ is right at the limiting value of $\simeq 0.07$.

5.3 Two-zero Mass Matrices

Of all the two-zero matrices classified in our previous work the only one to pass the constraints Eqs. (8) and (9) is the matrix N_3 in Table I. Two different roots (B_2 and B_3) produced matrices in agreement with Eqs. (8) and (9). However, $N_3(B_2)$ does not satisfy the new constraints we have applied, therefore, it will not be discussed further here.

5.3.1 $N_3(B_3)$

$N_3(B_3)$ (Figs. 16–18) is the only matrix that allows a low t-quark mass, $m_t \gtrsim 35$ GeV. It should be noted, however, that for low m_t values a fairly large value of the "bag" parameter is required ($B \gtrsim 1.2$). Also, we should remember that the loose constraints which we are using are very conservative. If for some range of m_t values a matrix only satisfies the loose constraints, then within this range it should be considered somewhat suspect. The tight constraints are found to be satisfied for t-quark masses roughly in the range $m_t = 55 - 80$ GeV. From Fig. 18 it is seen that, as in the case of M_5, if the upper limit on $\mid V_{ub}/V_{cb} \mid$ is substantially lowered then this matrix would be eliminated from further consideration. As for the other matrices examined this matrix also yields reasonable values for ϵ'/ϵ.

6 CONCLUSIONS

In this talk we have studied the consequences of imposing constraints from CP violation, $B^o - \bar{B}^o$ mixing, and the charmless decays of B-mesons on KM matrices arising from Hermitian quark mass matrices. We find that the three generation Fritzsch model is effectively ruled out by current experimental constraints. We found several alternative three generation matrices which do satisfy present constraints. Of these only two $M_6(A_2)$ and $M_6(A_3)$ were found to satisfy our constraints for $m_t > 100$ GeV. All matrices except $M_6(A_3)$ satisfy at least the loose constraints for $m_t \simeq 55 - 80$ GeV. The current limits on the ratio $\mid V_{ub}/V_{cb} \mid$ were found to provide a strong constraint.

From this work it is clear that we need to improve our understanding of the quark mixing sector. The constraints placed on the SM by $B^o - \bar{B}^o$ mixing data are greatly weakened by the uncertainties in f_B, B_B, and τ_B. Improvements in the calculation of ϵ and ϵ'/ϵ would greatly constrain the SM. The current experimental limits on these parameters are quite good but could also be improved. If the theoretical ambiguities involved in calculating x_d, ϵ, and ϵ'/ϵ can be reduced then it may be possible uniquely to determine the correct form of the quark mass matrix.

References

[1] Weinberg, S., Phys. Rev. Lett. 19, 1264 (1967); Phys. Rev. D5, 1412 (1972); Salam, A., in Elementary Particle Theory: Relativistic Groups and Analyticity (Nobel Symposium No. 8), edited by N. Svartholm (Almquist and Wiksell, Stockholm, 1968), p. 367; Glashow, S., Iliopolous, J., and Maiani, L., Phys. Rev. D2, 1285 (1970).

[2] Kobayashi, M., and Maskawa, T., Prog. Theor. Phys. 49, 652 (1973).

[3] Fritzsch, H., Nucl. Phys. B155, 189 (1979), Phys. Lett. 73B 317 (1978); Li, L.F., ibid 84B, 461 (1979).

[4] See, for example, Shin, M., Phys. Lett. 145B, 285 (1984); Harvard University Report No. HUTP-84/A070, 1984(unpublished); Georgi, H., Nelson, A., and Shin, A., Phys. Lett. 150B, 306 (1985); Georgi, H., and Nanopoulos, D.V., Nucl. Phys. B155, 52 (1979); Cheng, T.P., and Li, L.F., Phys. Rev. D34, 219 (1986); Gronau, M., Johnson, R., and Schecter, J., Phys;. Rev. Lett. 54, 2176 (1985).

[5] Harari, H., and Nir, Y., SLAC Report No. SLAC-PUB-4341, 1987; Nir, Y., SLAC Report No. SLAC-PUB-4368, 1987.

[6] ARGUS Collaboration, Albrecht, H. et al., Phys. Lett. 192B, 245 (1987).

[7] Schmidt-Parzefall, W., invited talk at the 1987 International Symposium on Lepton and Photon Interactions at High Energies, Hamburg, 1987.

[8] Robinson, J.A., and Rizzo, T.G., Phys. Rev. D36, 885 (1987).

[9] Robinson, J.A., and Rizzo, T.G., Ames Laboratory Report No. IS-J 2734, 1987.

[10] Gasser, J., and Leutwyler, H., Phys. Rep. 87, 77 (1982).

[11] Amaldi, U. et al., Phys. Rev. D36, 1385 (1987).

[12] UA2 Collaboration, CERN Report No. CERN-EP/87-05 (1987).

[13] Rizzo, T.G., Mod. Phys. Lett. A2, 505 (1987).

[14] Halzen, F., Kim, C.S., and Willenbrock, S., University of Wisconsin, Madison Report No. MAD/PH/342.

[15] Althoff, M. et al., (TASSO Collaboration), Phys. Lett. 154B, 236 (1985); Adeva, B. et al., (Mark J Callaboration), ibid 152B, 439 (1985); Bartel, W. et al., (JADE Collaboration), ibid 160B, 337 (1985); Behrend, H.J. et al., (CELLO Collaboration), ibid 144B, 297 (1984).

[16] Adachi, I. et al., (TOPAZ Collaboration), Phys. Rev. Lett. 60, 97 (1988).

[17] Wimpenny, S., invited talk at the International Symposium on The Production and Decay of Heavy Flavors, Stanford, CA., 1987.

[18] Marciano, W., invited talk at the International Symposium for the Fourth Family of Quarks and Leptons, Santa Monica, CA., 1987.

[19] Behrends, S. et al., (CLEO Collaboration), Phys. Rev. Lett. 59, 407 (1987).

[20] Inami, T., and Lim, C.S., Prog. Theor. Phys. 65, 297 (1981).

[21] Particle Data Group, Phys. Lett. 170B, (1986).

[22] Gilman, F.J., and Wise, M.B., Phys. Rev. D27, 1128 (1983).

[23] Gaillard, M.K., and Lee, B.W., Phys. Rev. D10, 897 (1974).

[24] Reinders, L.J., and Yazaki, S., Nucl. Phys. B288, 789 (1987).

[25] See, for example, the discussion by Buras, A.J., in Proceedings of the International Europhysics Conference on High Energy Physics, Bari, Italy, 1985, edited by Nitti, L., and Preparata, G.

[26] Black, J.K. et al., Phys. Rev. Lett. 54, 1628 (1985).

[27] Bernstein, R.H. et al., Phys. Rev. Lett. 54, 1631 (1985).

[28] Winstein, B., invited talk at the Division of Particles and Fields Meeting, Salt Lake City, Utah, 1987.

[29] Manneli, I., invited talk at the 1987 International Symposium on Lepton and Photon Interactions at High Energies, Hamburg, 1987.

[30] Buras, A.J., and Gerard, J.M., Phys. Lett. 192B, 156 (1987).

[31] Donoghue, J.F., Golowich, E., and Holstein, B.R., Phys. Lett. 179B, 361 (1986).

[32] For example, see Ellis, J., Hagelin, J.S., and Rudaz, S., Phys. Lett. 192B, 201 (1987); Narison, S., CERN Report No. CERN-TH.4768/87, 1987; Shifman, M., invited talk at the 1987 International Symposium on Lepton and Photon Interactions at High Energies, Hamburg, 1987.

[33] This is the same conclusion reached in [Ref. 5] where the constraint on $| V_{ub}/V_{cb} |$ from charmless B-Decays was not included.

Table I: Hermitian quark mass matrices which satisfy the constraints in Eqs. (8), (9), (10), (13a) and (20a).

$$M_F = \begin{bmatrix} 0 & A & 0 \\ A & 0 & B \\ 0 & B & C \end{bmatrix} \quad M_5 = \begin{bmatrix} A & A & 0 \\ A & C & B \\ 0 & B & 0 \end{bmatrix} \quad M_6 = \begin{bmatrix} C & A & 0 \\ A & A & B \\ 0 & B & 0 \end{bmatrix}$$

$$M_9 = \begin{bmatrix} 0 & A & 0 \\ A & A & B \\ 0 & B & C \end{bmatrix} \quad N_3 = \begin{bmatrix} 0 & A & B \\ A & 0 & A \\ B & A & C \end{bmatrix}$$

Table II: Values of ϵ'/ϵ predicted by mass matrices satisfying the constraints in Eqs. (8), (9), (10), (13a) and (20a).

	m_t (GeV)					
	50	60	70	80	100	180
M_5	3.5 –3.5	0.95–3.2	1.1 –2.6	—	—	—
$M_6(A_2)$	1.4 –1.7	1.2 –1.5	1.1 –1.4	1.0 –1.3	0.94–1.2	0.74–1.1
$M_6(A_3)$	–	–	1.2 –1.6	0.95–1.4	0.85–1.4	0.83–1.2
$M_9(A_3)$	–	1.2 –1.6	0.98–1.5	1.4	—	—
$N_3(B_3)$	0.98–1.0	0.80–1.6	0.70–1.5	—	—	—

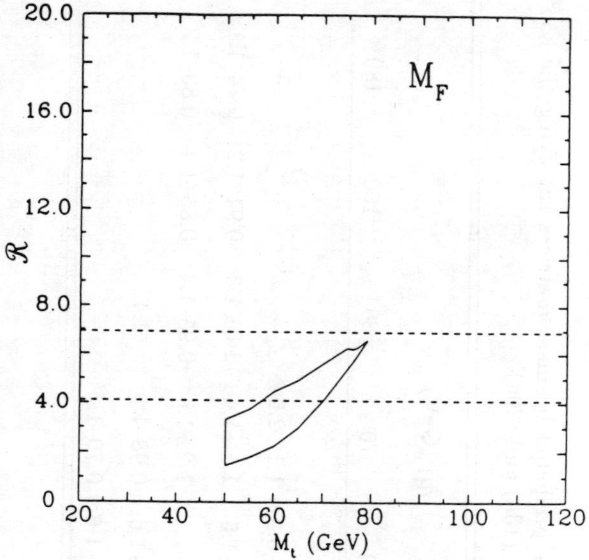

Fig. 1: The ratio R (defined in Eq. (19)) as a function of the t-quark mass for the matrix M_F. These values for R are obtained from KM matrices satisfying the constraints (8) and (9).

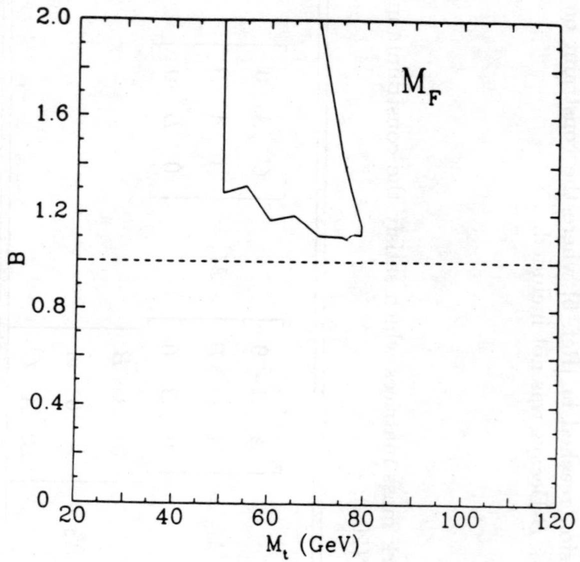

Fig. 2: B as a function of the t-quark mass for the matrix M_F. These values for B are obtained from KM matrices satisfying the constraints (8) and (9).

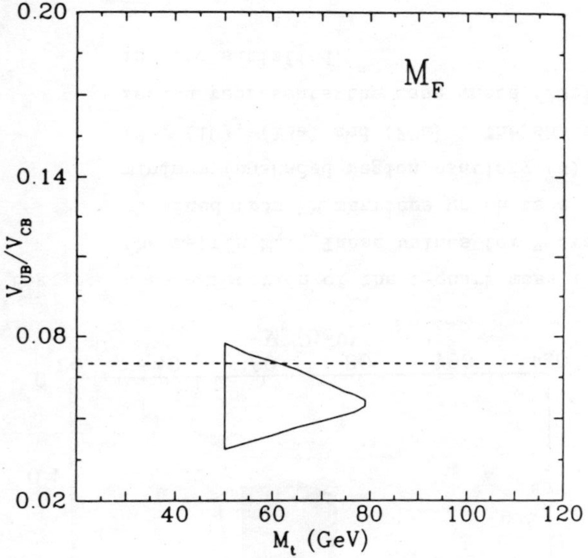

Fig. 3: $|V_{ub}/V_{cb}|$ as a function of the t-quark mass for the matrix M_F. Values for this ratio are obtained from KM matrices satisfying the constraints (8) and (9).

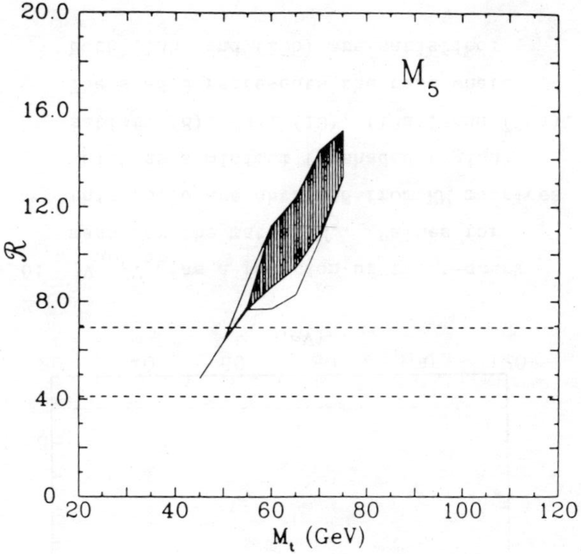

Fig. 4: The ratio R (defined in Eq. (19)) as a function of the t-quark mass for the matrix M_5. These values for R are obtained from KM matrices which as a minimum (unshaded region) satisfy (8), (9), (10), (13a), and (20a). The shaded region represents the case where (13b) is also satisfied

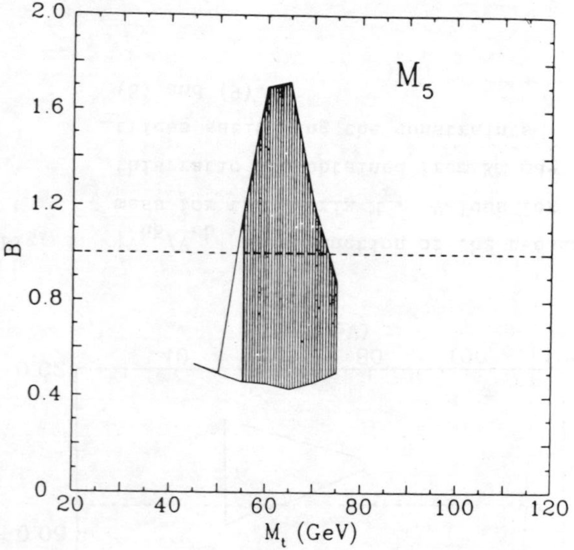

Fig. 5: B as a function of the t-quark mass for the matrix M_5. These values for B are obtained from KM matrices which as a minimum (unshaded region) satisfy (8), (9), (10), (13a) and (20a). The shaded region represents the case where (20b) is also satisfied.

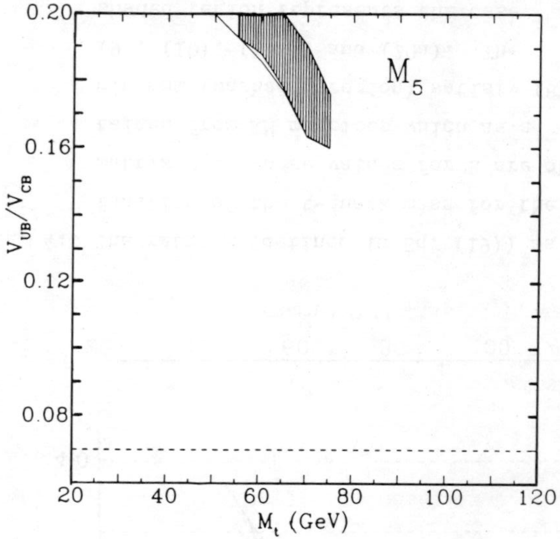

Fig. 6: $|V_{ub}/V_{cb}|$ as a function of the t-quark mass for the matrix M_5. Values for this ratio are obtained from KM matrices which as a minimum (unshaded region) satisfy (8), (9), (10), (13a), and (20a). The shaded represents the case where both (13b) and (20b) are satisfied.

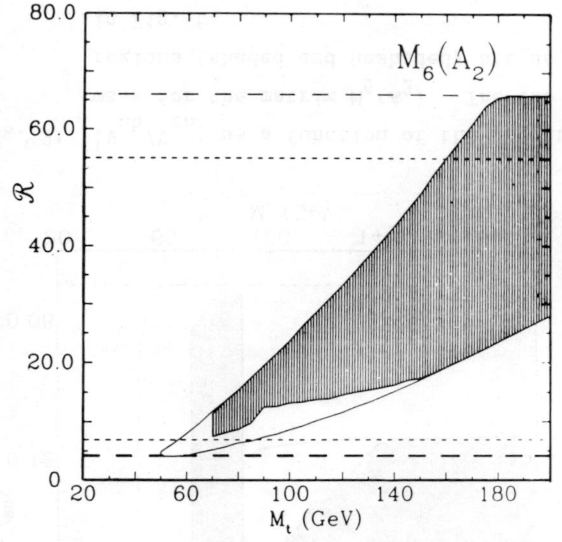

Fig. 7: The ratio R as a function of the t-quark mass for the matrix $M_6(A_2)$. The two regions (shaded and unshaded) are as in Fig. 4.

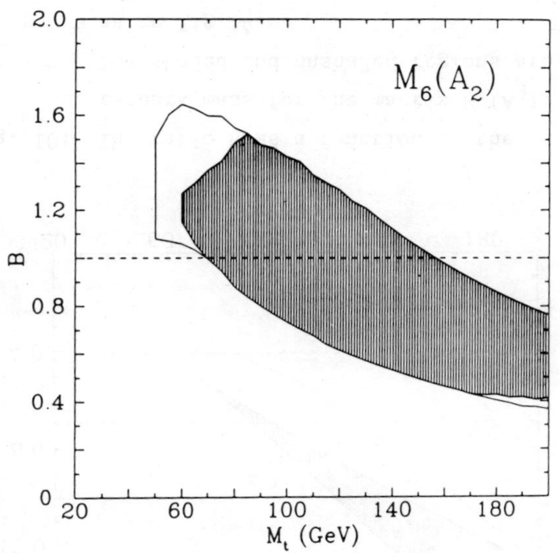

Fig. 8: B as a function of the t-quark mass for the matrix $M_6(A_2)$. The two regions (shaded and unshaded) are as in Fig. 5.

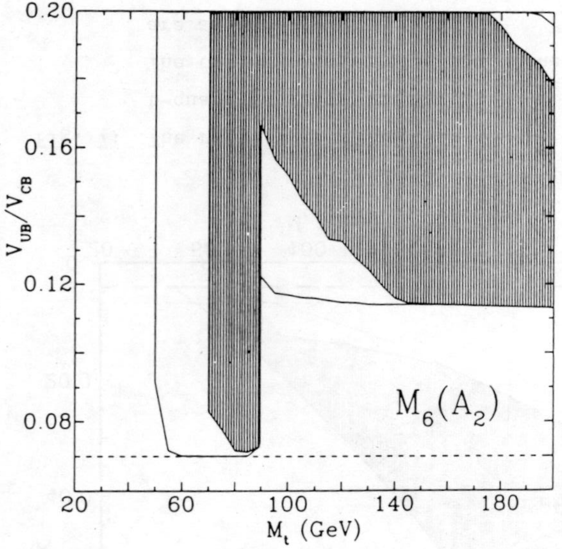

Fig. 9: $|V_{ub}/V_{cb}|$ as a function of the t-quark mass for the matrix $M_6(A_2)$. The two regions (shaded and unshaded) are as in Fig. 6.

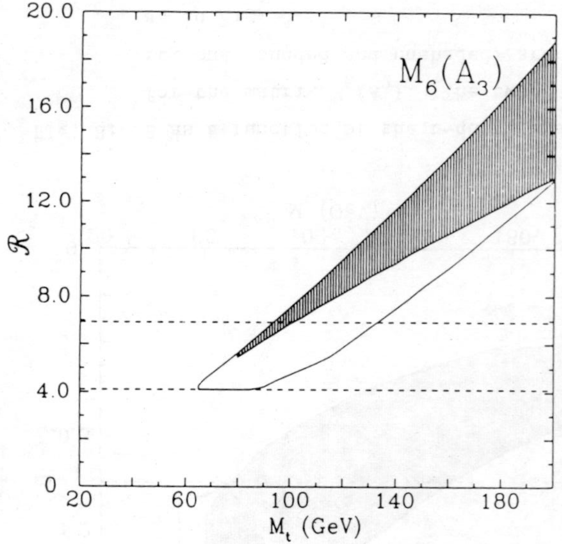

Fig. 10: The ratio R as a function of the t-quark mass for the matrix $M_6(A_3)$. The shaded and unshaded regions are as in Fig. 4.

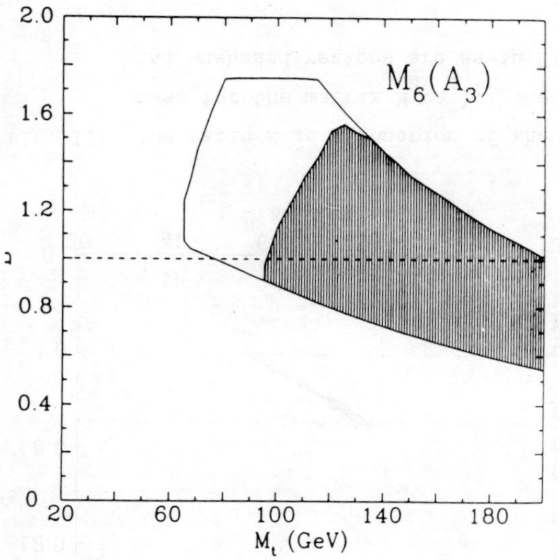

Fig. 11: B as a function of the t-quark mass for the matrix $M_6(A_3)$. The shaded and unshaded regions are as in Fig. 5.

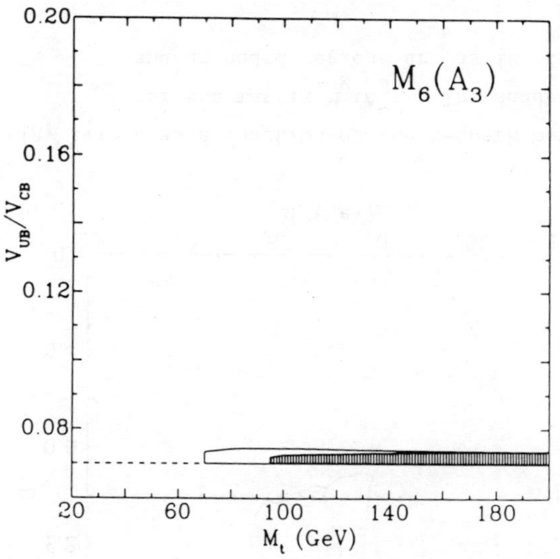

Fig. 12: $|V_{ub}/V_{cb}|$ as a function of the t-quark mass for the matrix $M_6(A_3)$. The shaded and unshaded regions are as in Fig. 6.

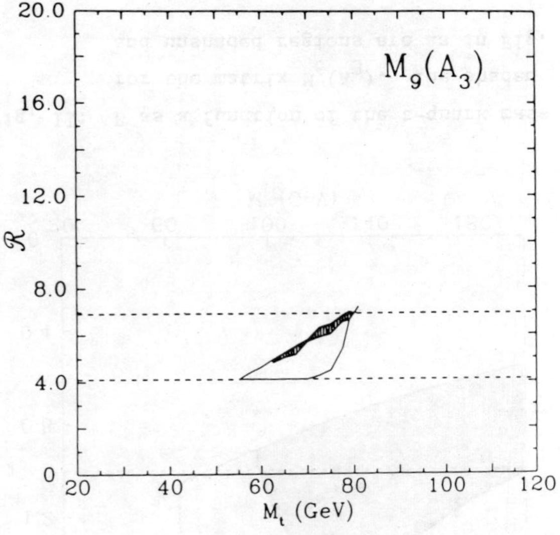

Fig. 13: The ratio R as a function of the t-quark mass for the matrix $M_g(A_3)$. The shaded and unshaded regions are as in Fig. 4.

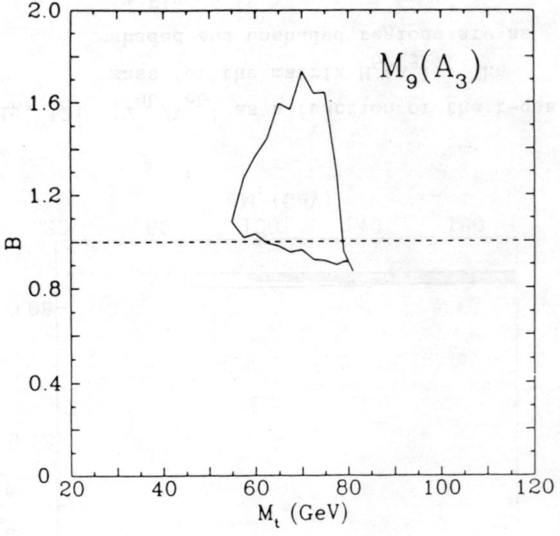

Fig. 14: B as a function of the t-quark mass for the matrix $M_g(A_3)$. The shaded and unshaded regions are as in Fig.5.

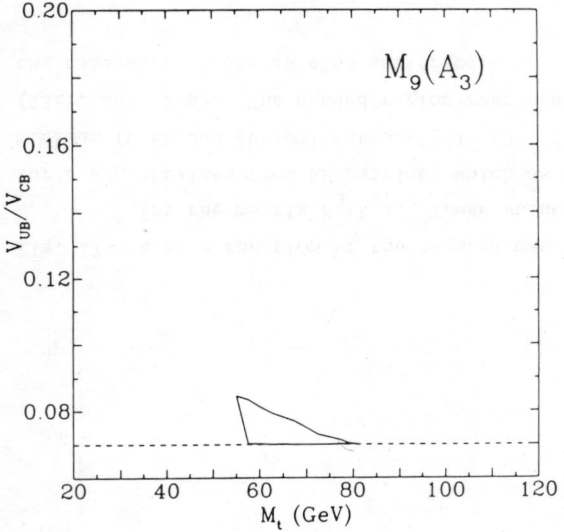

Fig. 15: $|V_{ub}/V_{cb}|$ as a function of the t-quark mass for the matrix $M_g(A_3)$. The shaded and unshaded regions are as in Fig. 6.

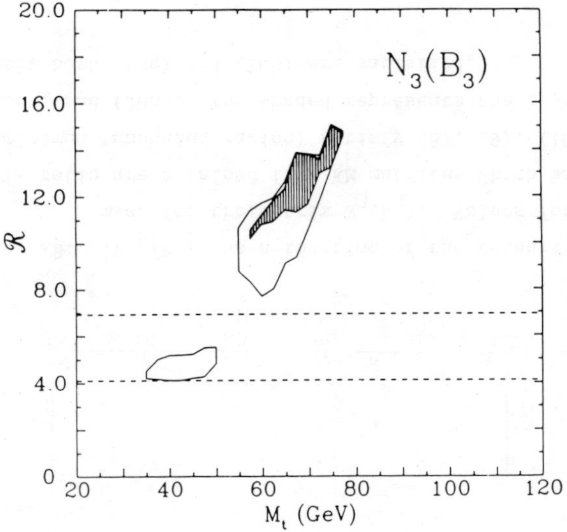

Fig. 16: The ratio R as a function of the t-quark mass for the matrix $N_3(B_3)$.

These values for R are obtained from KM matrices which as a minimum (unshaded region) satisfy (8), (9), (10), (13a), and (20a). The shaded region represents the case where (13b) is also satisfied.

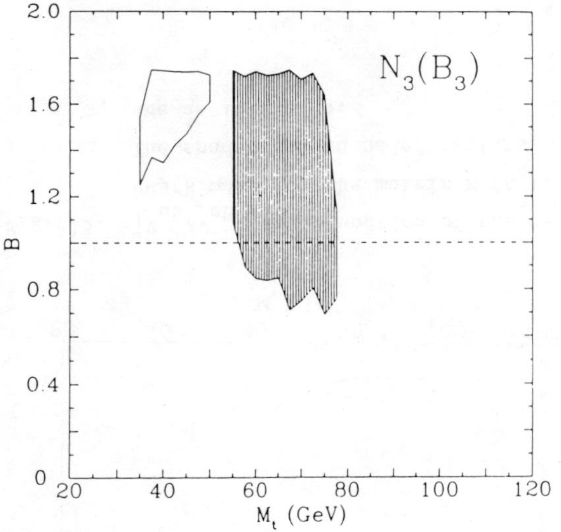

Fig. 17: B as a function of the t-quark mass for the matrix $N_3(B_3)$. These values for B are obtained from KM matrices which as a minimum (unshaded region) satisfy (8), (9), (10), (13a), and (20a). The shaded region represents the case where (20b) is also satisfied.

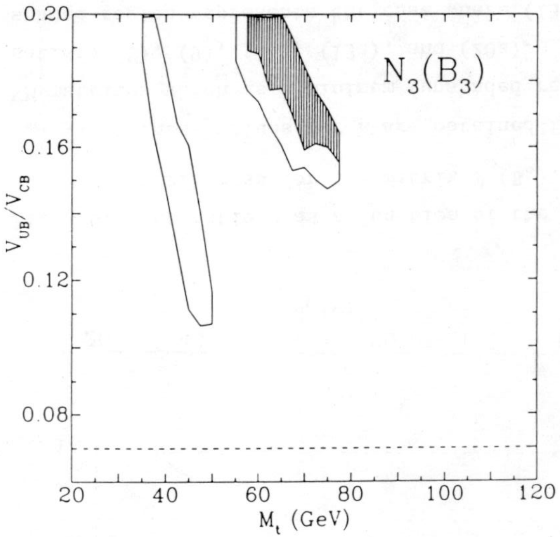

Fig. 18: $|V_{ub}/V_{cb}|$ as a function of the t-quark mass for the matrix $N_3(B_3)$. Values for this ratio are obtained from KM matrices which as a minimum (unshaded region) satisfy (8), (9), (10), (13a), and (20a). The shaded represents the case where both (13b) and (20b) are satisfied.

NOTES ON B-MESON MIXING AND DECAY ASYMMETRIES

Serge Rudaz

School of Physics and Astronomy
University of Minnesota
Minneapolis, MN 55455, USA

ABSTRACT

This article presents an informal and discursive theoretical overview of B^o-$\bar{B}^o$ mixing as evidenced in like-sign dilepton signals from the semileptonic decays of $B^o_d \bar{B}^o_d$ pairs produced at the Υ(4s) resonance in e^+e^- annihilation. The implications of the observation of such a signal by the ARGUS Collaboration as regards the Standard Model are quickly outlined, as are the prospects for the search for CP-non invariance in B-meson studies.

1. WEAK MIXING OF NEUTRAL PSEUDOSCALAR MESONS: GENERAL FORMALISM

We shall denote the neutral strange, charmed and bottom pseudoscalar mesons and their antiparticles, eigenstates of flavour and of $H_o = H_{st} + H_{em}$ by kets $|P^o\rangle$ and $|\bar{P}^o\rangle$. As a result of second order weak Δ(flavour) = ΔQ weak transitions $P^o \rightleftarrows$ "n" $\rightleftarrows \bar{P}^o$ into real or virtual states "n", the H_o eigenstates mix. This mixing will be described by the simple two-state effective Hamiltonian H_{eff}

$$H_{eff} = \mathcal{M}_{11}\, |P^o\rangle \langle P^o| + \mathcal{M}_{12} |P^o\rangle \langle \bar{P}^o$$

$$+ \mathcal{M}_{21}\, |\bar{P}^o\rangle \langle P^o| + \mathcal{M}_{22} |\bar{P}^o\rangle \langle \bar{P}^o|$$

$$\neq H^{\dagger}_{eff} \qquad (1)$$

Note H_{eff} is not hermitian, as the P^o and $\bar{P}^o$ are unstable: From CPT invariance only and hermiticity of H_o and H_{wk}, one has the constraint

$$\mathcal{M}_{11} = \mathcal{M}_{22} \qquad (2)$$

The assumption of T invariance alone implies

$$\mathcal{M}_{12} = e^{i\alpha_T} \mathcal{M}_{21} \qquad (3)$$

where the phase α_T is arbitrary: a redefinition of the relative phase of $|P^o\rangle$ and $|\bar{P}^o\rangle$ (not an observable because H_o is rigourously flavor conserving), for example $|P^o\rangle \rightarrow |P^o\rangle$, $|\bar{P}^o\rangle \rightarrow e^{i\alpha_T/2} |\bar{P}^o\rangle$ then allows the T-invariance constraint to be written as

$$\mathcal{M}_{12} = \mathcal{M}_{21}$$

We shall only assume CPT invariance here: a phase convention independent indication of T-(i.e. CP-) non-invariance in H_{eff} would then be $|\mathcal{M}_{12}| \neq |\mathcal{M}_{21}|$.

The $\mathcal{M}_{ij}$ can be thought of as the elements of a general 2 x 2 matrix with complex entries: such a matrix $\underset{\sim}{\mathcal{M}}$ can always be decomposed into two hermitian matrices $\underset{\sim}{M}$ and $\underset{\sim}{\Gamma}$,

$$\underset{\sim}{\mathcal{M}} = \underset{\sim}{M} - i\,\underset{\sim}{\Gamma}/2 \tag{4}$$

with

$$\underset{\sim}{M} = \underset{\sim}{M}^\dagger, \quad \underset{\sim}{\Gamma} = \underset{\sim}{\Gamma}^\dagger \tag{5}$$

so quite generally one has that M_{11} and M_{22}, as well as Γ_{11} and Γ_{22} are all <u>real</u>, while $M_{21} = M_{12}^*$ and $\Gamma_{21} = \Gamma_{12}^*$. The CPT-invariance constraint $\mathcal{M}_{11} = \mathcal{M}_{22}$ then implies

$$\mathcal{M}_{11} = \mathcal{M}_{22} \equiv M - i\Gamma/2 \tag{6}$$

where $M = M_{11} = M_{22}$, $\Gamma = \Gamma_{11} = \Gamma_{22}$ are real. H_{eff} is easily diagonalized: one finds the eigenstates

$$|P_1\rangle = (1 + |\eta|^2)^{-1/2}\,(|P^o\rangle + \eta\,|\bar{P}^o\rangle)$$

$$|P_2\rangle = (1 + |\eta|^2)^{-1/2}\,(|P^o\rangle - \eta\,|\bar{P}^o\rangle) \tag{7}$$

where η is given by

$$\eta = (\mathcal{M}_{21}/\mathcal{M}_{12})^{1/2} \tag{8}$$

and with $\mathcal{M}_{21} = M_{12}^* - i\Gamma_{12}^*/2$, $\mathcal{M}_{12} = M_{12} - i\Gamma_{12}/2$. Note that unless CP invariance holds, the eigenstates $|P_1\rangle$ and $|P_2\rangle$ are not orthogonal,

$$\langle P_2|P_1\rangle = \frac{1 - |\eta|^2}{1 + |\eta|^2} \tag{9}$$

η itself is phase convention dependent, but $|\eta| \neq 1$ is a phase convention independent signal of CP-non-invariance in the mixing. The corresponding eigenvalues are $\lambda_i = m_i - i\Gamma_i/2$, $i = 1, 2$, where

$$M = \frac{1}{2}(m_1 + m_2),\ \Gamma = \frac{1}{2}(\Gamma_1 + \Gamma_2)$$

$$\Delta m = m_1 - m_2 = \text{Re}(\lambda_1 - \lambda_2) = 2\ \text{Re}(\mathcal{M}_{12}\mathcal{M}_{21})^{1/2} \tag{10}$$

$$\Delta\Gamma = \Gamma_1 - \Gamma_2 = -2\text{Im}(\lambda_1 - \lambda_2) = -4\text{Im}(\mathcal{M}_{12}\mathcal{M}_{21})^{1/2}$$

i.e.

$$\Delta m = 2\text{Re}(|M_{12}|^2 - |\Gamma_{12}|^2/4 - i\ \text{Re}\ M_{12}\Gamma_{12}^*)^{1/2}$$

$$\Delta\Gamma = -4\text{Im}(|M_{12}|^2 - |\Gamma_{12}|^2/4 - i\ \text{Re}\ M_{12}\Gamma_{12}^*)^{1/2} \tag{11}$$

Writing $\eta \equiv q/p$, one can express $|P^o\rangle$, $|\bar{P}^o\rangle$ in terms of the eigenstates of H_{eff},

$$|P^o\rangle = \frac{\mathcal{N}}{2p}(|P_1\rangle + |P_2\rangle)$$

$$|\bar{P}^o\rangle = \frac{\mathcal{N}}{2q}(|P_1\rangle - |P_2\rangle) \tag{12}$$

with $\mathcal{N} = (|p|^2 + |q|^2)^{1/2}$. These equations allow us to determine the proper time (t) evolution of a state prepared as an eigenstate of H_o at $t = 0$ through the simple time dependence of the eigenstates of H_{eff} in the Weisskopf-Wigner approximation, namely

$$|P_i(t)\rangle = e^{-i\lambda_i t}|P_i(t = 0)\rangle \tag{13}$$

Thus,

$$|P^o(t)\rangle = f_+(t)\ |P^o\rangle + \eta\ f_-(t)\ |\bar{P}^o\rangle$$

$$|\bar{P}^o(t)\rangle = f_+(t)\ |\bar{P}^o\rangle + \eta^{-1}\ f_-(t)\ |P^o\rangle \tag{14}$$

where by a standard abuse of notation $|P^o(t)\rangle$ represents the state at proper time t that has evolved from the state $|P^o\rangle$ at $t = 0$, and similarly for $|\bar{P}^o(t)\rangle$: Of course, $|P^o\rangle$ and $|\bar{P}^o\rangle$ are time-independent basis vectors. The functions $f_\pm(t)$ are given by

$$f_{\pm}(t) = 1/2\,(e^{i\lambda_1 t} \pm e^{i\lambda_2 t}) \tag{15}$$

The semileptonic decays of neutral pseudoscalar mesons provide a convenient tag for the $|P^o\rangle$ and $|\bar{P}^o\rangle$ components of an initially pure $|P^o\rangle$ or $|\bar{P}^o\rangle$ state at proper time t. Indeed, with the assignments

$$P^o = K^o(\bar{s}d),\ D^o(c\bar{u}),\ B_d^o\ (\bar{b}d),\ B_s^o(\bar{b}s)$$

one has the general Δ(flavour) = ΔQ rule for semi-leptonic decays

$$P^o \to \ell^+\nu_\ell + X \qquad \text{only}$$
$$\bar{P}^o \to \ell^-\bar{\nu}_\ell + X \qquad \text{only}$$

so a positive lepton tags the $|P^o\rangle$ component, and a negative one, the $|\bar{P}^o\rangle$ component. One easily finds the normalized decay probabilities as a function of proper time

$$N\,\frac{dP}{dt}\,[P^o(t) \to \ell^+ + X] = |f_+(t)|^2$$

$$N\,\frac{dP}{dt}\,[P^o(t) \to \ell^- + X] = |\eta|^2|f_-(t)|^2$$

$$\bar{N}\,\frac{dP}{dt}\,[\bar{P}^o(t) \to \ell^- + X] = |f_+(t)|^2 \tag{16}$$

$$\bar{N}\,\frac{dP}{dt}\,[\bar{P}^o(t) \to \ell^+ + X] = |\eta|^{-2}|f_-(t)|^2$$

where N and $\bar{N}$ are calculated from the normalization

$$\int_o^\infty dt\,\frac{dP}{dt}\,[\ell^+, \ell^-] = 1$$

up to a branching ratio factor. If $|\eta| = 1$, it is easy to determine N $= \bar{N} = (\Gamma_1^{-1} + \Gamma_2^{-1})/2 = \tau$ to be the average lifetime of the eigenstates of H_{eff}: This follows from

$$|f_{\pm}(t)|^2 = 1/2e^{-\Gamma t}\,(\cosh\Delta\Gamma t/2 \pm \cos\Delta mt) \tag{17}$$

and from the integral

$$\int_0^\infty dt|f_\pm(t)|^2 = \frac{\Gamma}{2}\left[\frac{1}{\Gamma^2-(\Delta\Gamma)^2/4} \pm \frac{1}{(\Delta m)^2+\Gamma^2}\right] \tag{18}$$

with Γ, ΔΓ and Δm defined in Equation (10).

2. WEAK MIXING: OBSERVATIONS

The relatively long lifetimes of both $|K_L>$ $(=|K_1>)$ and $|K_S>$ $(=|K_2>)$ allows for the direct observation of K^o-$\bar{K}^o$ oscillations by a variety of means. The long lived kaon K_L decays mainly to three pions, while the short lived one K_S mainly to two pions, leading to $\Gamma_S >> \Gamma_L$ mainly because of phase space effects: The relevant parameters for the kaon system are determined to be

$$\frac{\Delta m}{\Delta\Gamma} \simeq -\frac{\Delta m}{\Gamma_S} = -0.477$$

$$\frac{\Delta\Gamma}{\Gamma} \simeq -2 \rightarrow \left|\frac{\Delta m}{\Gamma}\right|_K \simeq 1 \tag{19}$$

The first evidence for CP non-invariance in nature came with the 1964 discovery by Christenson, Cronin, Fitch and Turlay of the two charged pion decay mode of K_L with branching ratio 2×10^{-3}, indicating that $|K_L>$ could not be a CP-eigenstate. Subsequently, the decay $K_L \rightarrow \pi^o\pi^o$ was also established as further evidence for CP non-invariance. It is clear that there are two possible sources of CP non-invariance: in the $|\Delta S| = 2$ mixing, via $|\eta| \neq 1$, and in the $|\Delta S| = 1$ non-leptonic decay amplitude. To disentangle the two, consider the (until very recently, only) further example of CP-non-invariance in the neutral kaon system, namely the observation of a semileptonic decay asymmetry for K_L,

$$\delta_\ell = \frac{\Gamma(K_L\rightarrow \pi^-\ell^+\nu_\ell) - \Gamma(K_L\rightarrow \pi^+\ell^-\bar{\nu}_\ell)}{\Gamma(K_L\rightarrow \pi^-\ell^+\nu_\ell) + \Gamma(K_L\rightarrow \pi^+\ell^-\bar{\nu}_\ell)} = 3.30 \times 10^{-3} \tag{20}$$

an average over experiments with $\ell = e, \mu$. Given the $\Delta S = \Delta Q$ rule, δ_ℓ is easily calculated from the above formalism since only the K^o components can lead to ℓ^+ and the $\bar{K}^o$ to ℓ^-: one finds

$$\delta_\ell = \frac{|A_{s\ell}|^2 - |\eta|^2 \, |\bar{A}_{s\ell}|^2}{|A_{s\ell}|^2 + |\eta|^2 \, |\bar{A}_{s\ell}|^2} \tag{21}$$

where the semileptonic decay amplitudes are given by

$$A_{s\ell} = \langle \pi^- \ell^+ \nu_\ell | H_{wk} | K^o \rangle$$

$$\bar{A}_{s\ell} = \langle \pi^+ \ell^- \bar{\nu}_\ell | H_{wk} | \bar{K}^o \rangle \tag{22}$$

Assuming CPT invariance and hermiticity of H_{wk} implies $|A_{s\ell}| = |\bar{A}_{s\ell}|$, so

$$\delta_\ell = \frac{1 - |\eta|^2}{1 + |\eta|^2} \tag{23}$$

and the semileptonic decay asymmetry is a direct measurement of CP-non invariance in the mixing of K^o and $\bar{K}^o$ only. As a result, one extracts

$$|\eta_K|^2 = 1 - \frac{\mathrm{Im}\, \Gamma_{12}/M_{12}}{1 + |\Gamma_{12}|^2/4|M_{12}|^2}$$

$$= \frac{1 - \delta_\ell}{1 + \delta_\ell} \simeq 1 - 2\delta_\ell \tag{24}$$

The fact that $\delta_\ell \ll 1$ means that to a very good accuracy one can write the simplified relations

$$\Delta m \simeq 2|M_{12}|$$

$$\Delta\Gamma \simeq -2|\Gamma_{12}| \tag{25}$$

(the signs are fixed by data) and using $\Delta m/\Delta\Gamma \simeq -1/2$ one determines

$$\mathrm{Im}(\Gamma_{12}/M_{12})_K \simeq 1.3 \times 10^{-2} \tag{26}$$

as a phase convention independent characterization of CP violation in the $|\Delta S| = 2$ mixing amplitude for the $K^o - \bar{K}^o$ system. Additionally, and very recently, the first experimental evidence for CP non-invariance in the $K \to \pi\pi$, $|\Delta S| = 1$ decay amplitudes has come from the measurement by the NA31 collaboration at CERN of a non-vanishing value of the famous ratio

$$\mathrm{Re}\ \epsilon'/\epsilon_o \simeq \frac{1}{6}\,(1 - |\eta_{oo}|^2/|\eta_{+-}|^2) = (3.3 \pm 1.1) \times 10^{-3} \tag{27}$$

where $\eta_{+-} = A(K_L \to \pi^+\pi^-)/A(K_S \to \pi^+\pi^-)$ and analogously for η_{oo} for the neutral pion decay mode. We will have only a little to say about the relevance of this result to fixing the parameters of the standard model below.

We now turn to a discussion of how $B_d^o - \bar{B}_d^o$ mixing was discovered by the ARGUS collaboration. The short lifetimes, of order a picosecond, of the B-mesons preclude direct observations of flavour oscillations (for the time being, at least). However, time integrated quantities that are sensitive to mixing do exist: for example, starting from a pure B_d^o state, we may calculate the relative integrated yields of "wrong sign" as compared to "right sign" leptons form semileptonic decays,

$$\frac{N("B_d^o" \to \ell^-)}{N("B_d^o" \to \ell^+)} = |\eta|^2\ \gamma \tag{28}$$

where

$$\gamma = \int_o^\infty dt\ |f_-(t)|^2 \Big/ \int_o^\infty dt\, |f_+(t)|^2$$

$$= \frac{(\Delta m/\Gamma)^2 + (\Delta\Gamma/2\Gamma)^2}{2 + (\Delta m/\Gamma)^2 - (\Delta\Gamma/2\Gamma)^2} \tag{29}$$

is easily obtained from Equation (18). Similarly, starting from a pure $\bar{B}^o_d$ state, one finds

$$\frac{N("\bar{B}^o_d" \to \ell^+)}{N("\bar{B}^o_d" \to \ell^-)} = |\eta|^{-2}\ \gamma \tag{30}$$

For an incoherently produced $B^o_d\ \bar{B}^o_d$ pair, say in $p\bar{p} \to B^o_d\ \bar{B}^o_d$ + anything, or $e^+e^- \to B^o_d\ \bar{B}^o_d$ + anything, one can similarly consider the relative yield of wrong sign dileptons arising form the semileptonic decays of both time evolved B-mesons: Thus,

$$\left.\frac{N(\ell^+\ell^+)+N(\ell^-\ell^-)}{N(\ell^+\ell^-)}\right|_{\text{incoherent}} =$$

$$= \frac{N("B^o_d" \to \ell^+)N("\bar{B}^o_d" \to \ell^+)+N("B^o_d" \to \ell^-)N("\bar{B}^o_d" \to \ell-)}{N("B^o_d" \to \ell^+)N("\bar{B}^o_d" \to \ell^-)+N("B^o_d" \to \ell^-)N("\bar{B}^o_d" \to \ell^+)}$$

$$= (|\eta|^2 + |\eta|^{-2})\ \frac{\gamma}{1+\gamma^2} \tag{31}$$

The situation is quite different if the initial $B^o_d\ \bar{B}^o_d$ state is prepared coherently in a single quantum state, as is the case for those pairs coming from the decay $\Upsilon(4s) \to B^o_d\ \bar{B}^o_d$. The B-meson pair is produced in the C and P conserving strong decay of the upsilon with $J^{PC} = 1^{--}$, and so is initially in a pure odd-C state

$$|\psi(\tau{=}0)>_{odd} = \frac{1}{\sqrt{2}}[\,|B^o_d;+>\ |\bar{B}^o_d;->\ -\ |B^o_d;->\ |\bar{B}^o_d;+>]$$

$$= -\ \frac{\mathcal{N}}{2\sqrt{2pq}}\ \ [\,|B_1;+>|B_2;->\ -\ |B_1;->|B_2;+>] \tag{32}$$

where "+" and "-" indicate a "right"-moving particle (respectively, "left"-moving) with respect, say, to some arbitrarily chosen plane. In the lab (CM) frame the pair is produced symmetrically, back-to-back and the eigenstates $|B_{1,2}>$ evolve along common proper times that are simply related to the lab time by the usual special relativistic factor. Thus, after a proper time t one has

$$|\psi(t)\rangle_{odd} = e^{-i(\lambda_1 + \lambda_2)t} |\psi(t=0)\rangle_{odd} \tag{33}$$

so in particular the B_d^o and $\bar{B}_d^o$ components maintain their original phase relation (this would <u>not</u> be the case had we started with the C-even eigenstate). Thus, an experiment seeking to determining whether the left-moving particle is a B_d^o at time t (for example, by observing a semileptonic decay into ℓ^+) will find it to be so with probability 1/2 (times exp $(-t/\tau_1 - t/\tau_2)$ because of the instability of the B eigenstates); however, once the left-moving particle has been identified at time t to be a B_d^o, then, <u>at the same time t</u>, the right-moving particle is determined to be a $\bar{B}_d^o$ <u>with absolute certainty</u> (and vice versa). This is an example of an Einstein-Podolsky-Rosen correlation ("spukhafte fernwirkungen" according to Einstein) and is an unambiguous prediction of quantum mechanics. It is easiest to understand by recalling that for a system of two charge conjugate pseudoscalar mesons, $C = P = (-1)^L$ so odd C implies odd orbital angular momentum: now Bose Einstein statistics forbids two identical bosons to be in an antisymmetric state, so at <u>a given time t</u>, $B_d^o B_d^o$ and $\bar{B}_d^o \bar{B}_d^o$ are forbidden configurations for odd L. Again, this is not so for the even C, even L initial $B^o \bar{B}^o$ configuration. These considerations allow for a simple derivation of the expression for the number of like-sign as compared to unlike-sign dileptons in the double semileptonic decays of $B_d^o \bar{B}_d^o$ pairs produced in $\Upsilon(4s)$ decays, taking due account of the quantum mechanical aspects of the measurement. One has

$$r_d \equiv \left. \frac{N(\ell^+\ell^+) + N(\ell^-\ell^-)}{N(\ell^+\ell^-)} \right|_{\text{coherent, odd C}}$$

$$= \text{Prob}(\ell^+) \frac{P(\text{"}\bar{B}_d^o\text{"} \to \ell^+)}{P(\text{"}\bar{B}_d^o\text{"} \to \ell^-)} + \text{Prob}(\ell^-) \frac{P(\text{"}B_d^o\text{"} \to \ell^-)}{P(\text{"}B_d^o\text{"} \to \ell^+)} \tag{34}$$

where, for example, in the first term of this equation, Prob(ℓ^+), the a priori probability that a positively charged lepton (i.e. by definition, a B_d^o semileptonic decay) is observed first (namely, $\frac{1}{2}$: cf. Equations (32), (33)), is multiplied by the probability for the oppositely directed particle (necessarily a $\bar{B}_d^o$ at that time) to then yield a wrong-sign relative to a right-sign lepton in its subsequent time evolution and decay. The second term is understood, mutatis mutandis, in the same way. Finally, with Equations (28) and (30), one finds

$$r = \frac{1}{2}(|\eta|^{-2} + |\eta|^2)\gamma \tag{35}$$

with γ as given in Equation (29). The like-sign dilepton asymmetry from the same source is trivially written down in the same way

$$a_d = \left.\frac{N(\ell^+\ell^+) - N(\ell^-\ell^-)}{N(\ell^+\ell^+) + N(\ell^-\ell^-)}\right|_{\text{coherent - odd C}}$$

$$= \frac{|\eta|^{-2} - |\eta|^2}{|\eta|^{-2} + |\eta|^2} = \frac{1 - |\eta|^4}{1 + |\eta|^4} \tag{36}$$

A non-vanishing value of a_d would be an indication of T-non-invariance (CP-non-invariance) in B_d^o - $\bar{B}_d^o$ mixing. To the extent that one expects $\Delta\Gamma/\Gamma << 1$ and that $|\eta| \simeq 1$ to a very good approximation in the B_d^o - $\bar{B}_d^o$ system (as justified in the next section), one has the simple formula

$$r_d = \frac{(\Delta m/\Gamma)^2}{2 + (\Delta m/\Gamma)^2} \tag{37}$$

The ARGUS collaborations report a value $r_d = 0.23 \pm 0.07$ which translates into

$$\left|\frac{\Delta m}{\Gamma}\right|_{B_d} \simeq 0.73 \pm 0.14, \tag{38}$$

not all that different form the neutral kaon value given in Equation 19: Confirmation of this result by the CLEO Collaboration is eagerly awaited! In the following, we shall assume the conservative bound

$$r_d \geq 0.10$$
$$\left|\frac{\Delta m}{\Gamma}\right|_{B_d} \geq 0.45 \tag{39}$$

to determine the implications of the observation of B_d^o - $\bar{B}_d^o$ mixing as regards the parameters of the three-generation Standard Model.

3. B_d^o - $\bar{B}_d^o$ MIXING AND THE STANDARD MODEL

We expect that the dispersive part M_{12} of the matrix element $\mathcal{M}_{12}$ of the effective Hamiltonian for B_q^o - $\bar{B}_q^o$ mixing can reliably be calculated (within the Standard Model) from the amplitude, given by the usual box diagrams, for the process $b\bar{q} \to \bar{b}q$ proceeding via two-W exchange at short distances (large relative momenta of the intermediate quarks). As usual, neglecting external quark masses and momenta as compared to the mass of the heavy quark whose contribution dominates the box amplitude one determines a local effective four-fermion Lagrangian for the mixing of the form

$$\mathcal{L}_{\text{eff}} = -\frac{G_F^2 m_t^2}{16\pi^2}(V_{tq}^* V_{tb})^2 \eta_{QCD} F(m_t^2/M_W^2) \cdot$$
$$\cdot [\bar{q}\gamma^\mu(1-\gamma_5)b][\bar{q}\gamma_\mu(1-\gamma_5)b] \tag{40}$$

appropriate to B_q^o - $\bar{B}_q^o$ (q = d or s) mixing, for which the top quark ($t\bar{t}$ box) contribution dominates. Here, the function F(x) is given by

$$F(x) = \frac{1 - 11x/4 + x^2/4}{(1 - x)^2} - \frac{3\, x^2 \ell n x}{2(1-x)^3} \simeq \frac{3}{4}\, x^{-1/8} \tag{41}$$

F(x) has the limiting values 1, 3/4 and 1/4 for $x \ll 1$, $x = 1$ and $x \gg 1$ respectively: the last approximate equality is an easily remembered and fairly reasonable approximation to F(x) for the range of top quark masses

$$30\ \text{GeV} \le m_t \le 180\ \text{GeV} \tag{42}$$

where the lower limit corresponds to the lower bound on m_t obtained from the so-far unsuccessful searches for $t\bar{t}$ production in e^+e^- annihilation, and the upper limit is a typical upper bound that results form an analysis of neutral current data (essentially, that the ρ-parameter measuring the relative strength of neutral to charge current interactions at low momentum transfers is one to within a few percent: This provides a bound on the weak isodoublet splitting $(m_t^2 - m_b^2)/M_W^2$ in the three generation case). The V_{ij} are CKM matrix elements: for quick estimates, Wolfenstein's mnemonic is quite useful

$$V = \begin{array}{c} \\ u \\ c \\ t \end{array} \begin{array}{c} \begin{array}{ccc} d & s & b \end{array} \\ \begin{bmatrix} 1-\lambda^2/2 & \lambda & A\lambda^3(\rho - i\eta) \\ -\lambda & 1-\lambda^2/2 & A\lambda^2 \\ A\lambda^3(1-\rho-i\eta) & -A\lambda^2 & 1 \end{bmatrix} \end{array} \tag{43}$$

In this form, V, describing flavour mixing in charged current interactions, is unitary up to terms of fourth order in the small parameter $\lambda \simeq \sin\theta_c \simeq 0.23$. The notation has become standard: There should be no risk of confusion between the parameter η for which a non-zero value is required for CP violation in charged-current couplings (i.e. so that $V \neq V^*$), and the ratio $(\mathcal{M}_{21}/\mathcal{M}_{12})^{1/2}$ appearing in the diagonalization of the effective Hamiltonian for mixing. Finally, $\eta_{QCD} \simeq 0.8$ is a short distance QCD dressing correction which is in fact quite insensitive to the precise value of m_t in the range $30\ \text{GeV} \le m_t \le 180\ \text{GeV}$ mentioned above.

The challenge is of course to compute the matrix element leading to M_{12} namely

$$M_{12} = - \frac{1}{2M_B} <B_d^o| \mathcal{L}_{eff}(0)|\bar{B}_d^o> \quad (44)$$

where $H_{eff} = - \int d^3x \mathcal{L}_{eff}(x)$, and the factor of $(2M_B)^{-1}$ compensates for the usual choice made for the normalization of momentum eigenstates at rest, as compared to the states normalized to one in our earlier discussion of the general formalism for mixing. The uncertainty in the evaluation of the matrix element is parametrized in terms of the so-called "B-parameter", defined as follows:

$$<B_q^o|\bar{q}_\alpha\gamma^\mu(1-\gamma_5)b_\alpha \; \bar{q}_\beta\gamma_\mu(1-\gamma_5)b_\beta| \; \bar{B}_q^o>$$

$$= 2(1 + \tfrac{1}{3}) \; B_{Bq}|<0|\bar{q}_\alpha\gamma_\mu\gamma_5 b_\alpha|\bar{B}_q^o>|^2$$

$$= \frac{8}{3} \; f_{Bq}{}^2 \; B_{Bq} \; M_{Bq}^2 \quad (45)$$

If $B_B = 1$, this is just the result of the vacuum saturation approximation which essentially amounts to the valence quark approximation for the external B-mesons: The decay constant f_{Bq} is defined by the usual matrix element

$$<0| \; \bar{q}_\alpha(x)\gamma_\mu\gamma_5 b_\alpha(x) \; |\bar{B}_q^o(P)> = i \; P_\mu f_{Bq} e^{-iP\cdot x} \quad (46)$$

In Equation (45), the indices α and β (summed over) are SU(3) colour indices: The second line, with B=1, reflects the four possible ways the vacuum can be inserted, with two of these leading to non-colour singlet current matrix elements. A colour Fierz transformation brings these contributions into a colour singlet form, with weight 1/3 = 1/N (N is the number of colours) plus a color non-singlet piece with vanishing matrix elements, hence the total factor 1 + 1 + 1/N + 1/N = 2(1 + 1/N) = 8/3 when N=3. We thus have two unknown parameters f_B and B that depend on non-perturbative physics.

Q1: What is f_B? Note that the width for the leptonic decay of a charged pseudoscalar meson into a lepton and its antineutrino is just

$$\Gamma(P(D\bar{U}) \to \ell^- \bar{\nu}_\ell) = \frac{G_F^2}{8\pi} |V_{UD}|^2 f_P^2 M_P^3 \left(\frac{m_\ell^2}{M_P^2} \right) (1 - m_\ell^2/M_P^2)^2 \tag{47}$$

In this normalization, $f_\pi = 133$ MeV and $f_K = 160$ MeV. An unsuccessful search for the decay $D \to \mu\nu$ by the Mark III collaboration leads to the limit

$$f_D < 290 \text{ MeV} \tag{48}$$

One generally expects that up to small perturbative QCD corrections, for heavy pseudoscalars P = D, B the following scaling law holds

$$f_P \propto 1/\sqrt{M_P} \tag{49}$$

Using this would lead to the bound

$$f_B \lesssim 170 \text{ MeV} \tag{50}$$

Q2: What is B_B? The accepted value of the analogous quantity for B_K has widely fluctuated over time between .3 and 1.5 as new calculations kept appearing. In my view, the following old argument based on the 1/N expansion provides the most convincing estimate: note that to leading order as $N \to \infty$, the vacuum saturation approximation is in fact exact. Using the parametric expression $2(1+1/N)$ explained above one finds then

$$B = 2/(8/3) = 3/4 \tag{51}$$

which should hold equally well for K, D and B meson matrix elements. Corrections to this simple estimate have been computed by Bardeen, Buras, Gerard and collaborators to be about 10%: coincidentally, nearly the same number has been obtained through lattice gauge theory computer calculations.

To determine the constraint on m_t that arises from the ARGUS observations, we first need to list the following constraints on CKM matrix elements from B-decays:

- from semileptonic B-decay one has the conservative bound

$$\bar{R} = \frac{\Gamma(b \to u\ell\nu)}{\Gamma(b \to c\ell\nu)} \leq 0.09 \qquad (52)$$

which after correcting for a phase space factor leads to a bound on $|V_{ub}|^2/|V_{cb}|^2$ which translates to

$$\rho^2 + \eta^2 \leq 1 \qquad (53)$$

- from the B-lifetime measured at $\tau_B = 1.1 \times 10^{-12}$ sec, together with the measured semileptonic branching ratio $BR(b \to e\nu X) = 11\%$ which given the bound on $\bar{R}$ translates into a value for $|V_{cb}|^2$ indicating that roughly

$$A \simeq 0.9 \qquad (54)$$

With this information, and $|\Delta m|_B = 2|M_{12}|_B$ one calculates

$$\left|\frac{\Delta m}{\Gamma}\right|_B = 2\tau_B|M_{12}|_B = 0.14 \left(\frac{f_B}{f_\pi}\right)^2 \left(\frac{B_B}{0.75}\right)\left(\frac{m_t}{M_W}\right)^{7/4} [(1-\rho)^2+\eta^2] \qquad (55)$$

To get a lower bound on m_t given $|\Delta m/\Gamma|_B$ is simple: recall that $\rho^2 + \eta^2 \leq 1$ while $\eta^2 > 0$ for CP-non-invariance to occur at all so

$$(1-\rho)^2 + \eta^2 \leq 2(1-\rho) \leq 2(1+\sqrt{1-\eta^2}) < 4 \qquad (56)$$

and thus

$$\frac{m_t}{M_W} > 1.4 \left(\frac{f_\pi}{f_B}\right)^{8/7} \left(\frac{0.75}{B_B}\right)^{4/7} \left|\frac{\Delta m}{\Gamma}\right|_B^{4/7} \qquad (57)$$

We now argue that for B decays, $\Delta\Gamma/2\Gamma \ll 1$. This result is intuitively clear, as B_1 and B_2 have many common decay channels, where by contrast to the extent that CP-non invariant effects are small, K_L and K_S decay into different channels (3π and 2π, respectively), with very different phase space factors leading to $\Gamma_L \ll \Gamma_S$ and $|\Delta\Gamma|2\Gamma| \simeq 1$. To get an idea of the magnitude of $|\Delta\Gamma|_B = 2|\Gamma_{12}|_B$ we use the naive quark model and neglect quark masses other than m_b. Γ_{12}, the absorptive part of $\mathcal{M}_{12}$ is calculated from the box contributions by putting internal quark lines on the mass-shell, which leave only the $(u\bar{u})$, $(c\bar{c})$, $(u\bar{c})$ and $(c\bar{u})$ contributions as kinematically accessible from an initial B-meson. Denoting by $\lambda_U = V^*_{Ub}V_{Ud}$ the product of CKM matrix elements appearing in the box graphs (with U = u, c, t) one estimates in this way

$$\Gamma_{12} \simeq \frac{G_F^2}{8\pi} f_B^2 M_B^3 (\lambda_u^2 + \lambda_c^2 + 2\lambda_u\lambda_c) = \frac{G_F^2}{8\pi} f_B^2 M_B^3 \lambda_t^2 \tag{58}$$

In the second step we have used the unitarity of the CKM matrix $(V^+)_{bU}V_{Ud} = 0$ which means $\lambda_u + \lambda_c + \lambda_t = 0$ so $(\lambda_u + \lambda_c)^2 = \lambda_t^2$. In the first expression, the CKM angle factors correspond to the four channels above, f_B^2 comes from the evaluation of the matrix element and the rest from dimensional analysis (even the power of pi can be determined by elementary counting). On the other hand, in the same approximation,

$$2\Gamma \simeq \frac{9G_F^2}{192\pi^3} M_B^5 |V_{cb}|^2 \tag{59}$$

so $$\left|\frac{\Delta\Gamma}{2\Gamma}\right| \sim O(\pi^2) \sin^2\theta_c \left(\frac{f_B}{M_b}\right)^2 \tag{60}$$

which given the crudeness of the argument could amount to a number at most of order a few percent. Thus, Equation (37) is justified and we can now place a bound on m_t. Using the inequalities in Equations (39) and (50) and with $B_B = 0.75$ as a benchmark value, Equation (57) lead to

$$m_t > 56 \text{ GeV} \tag{61}$$

while using the central ARGUS value in Equation (38) with $f_B = f\pi$ leads to the more stringent bound

$$m_t > 97 \text{ GeV} \tag{62}$$

These estimates agree with the results of the many more detailed analyses that have appeared. Generally, larger values for f_B and B_B weaken the bound. While on the contrary, a more stringent bound on $\bar{R}$ (Equation (52)) means a more stringent lower bound on m_t. It is clear that including CP-non-invariant effects in this analysis can only strengthen the bounds on m_t, to the extent that $Im(M_{12})_K = 0$ can be translated into $\eta^2 > 0$. However, as Equation (56) shows this will generally not be a large effect: while the bound on η^2 depends on B_K and m_t (typically, the larger the latter two quantities, the smaller η^2), taking $\eta^2 \geq 1/4$ as suggested by data leads to a strengthening of the bounds by only 5 to 10%. The recent measurement of ϵ'/ϵ_o in the two pion decay of K_L and K_S by the NA31 collaboration at CERN does not really give any new constraint on m_t, mainly as a result of the many uncertainties involved in the estimates of the matrix elements of the Penquin operators that contribute to ϵ': nevertheless, to the extent that a firm lower bound exists on m_t, one can place an upper bound on ϵ'/ϵ_o since this ratio is a monotonously decreasing function of m_t (viz. it behaves roughly as $(\ell n\ m_t)/m_t^p$ where p is some effective power). The resulting upper bound depends on the assumed value for B_K and increases with decreasing B_K: for $B_K = 0.33$, one expects $|\epsilon'/\epsilon_o| \lesssim 1 \times 10^{-2}$. No new lower bounds arise from information from $B_d^o - \bar{B}_d^o$ mixing: this remains at $|\epsilon'/\epsilon_o| > 1 \times 10^{-3}$ or so.

To conclude this section we see that these considerations leads us to expect the top quark be rather heavy, 60 GeV $\leq m_t \leq$ 180 GeV which would make its observation a marginal proposition even at the highest Fermilab Tevatron Collider or CERN-ACOL energies, and eliminate LEP I from contention for top quark studies. Some important immediate goals for the experiment, all of them extremely challenging, are as follows:

- a firm measurement of (or a firm refinement of the bound on) the ratio $\bar{R}$ determining the magnitude of $|V_{ub}|^2$;

- a measurement of the mixing parameter for the B_s^o - $\bar{B}_s^o$ system, which requires a tagging method to ensure that one is indeed seeing B_s^o decays: the Standard model leads to the expectation

$$|\Delta m/\Gamma|_{B_s}/|\Delta m/\Gamma|_{B_d} \simeq |V_{ts}|^2/|V_{td}|^2 > 4 \qquad (63)$$

where we have used the inequality (56).

- a measurement of f_B via the leptonic decay $B_u \to \tau\nu_\tau$: using Equations (47), (50) (53) and (54) one finds

$$BR(B_u \to \tau\nu_\tau) \le 4.2 \times 10^{-4} \qquad (64)$$

Because of the m_ℓ^2 factor in Equation (47), decays into electrons and muons are minuscule. Assuming a branching ratio of 10^{-4} for $B_u^- \to \tau\bar{\nu}_\tau$ decay, an optimistic tagging efficiency of, say, 10% for the accompanying B_u^+ and a 100% efficiency in τ-identification requires at least a million $B_u^+ B_u^-$ pairs for the identification of ten leptonic decay events! This appears to be a job for a B-factory: determining f_B must be considered a priority.

- a measurement of the branching ratio for the rare decay $K^+ \to \pi^+ \nu\bar{\nu}$ which is predicted to be in the range (1-8) x 10^{-10} for the range of m_t given above: an experiment is underway at Brookhaven that is predicted to eventually reach this sensitivity.

4. CP-NON INVARIANCE IN B-MESON PHYSICS

An obvious signal for CP-non invariance in neutral B_d^o - $\bar{B}_d^o$ mixing would be the observation of a non-vanishing same sign dilepton asymmetry at the Υ(4s). This asymmetry was defined in Equation (36) and it is easy to obtain, with $|\Delta\Gamma/\Delta m| << 1$, the expression (cf. Equation (24))

$$a_d \simeq \mathrm{Im}(\Gamma_{12}/M_{12})_{B_d} \qquad (65)$$

Our naive quark model estimate of Γ_{12} (neglecting phase space factors, which amounts to setting $m_u^2 = m_c^2 << m_b^2$) together with the unitarity of the three generation CKM matrix led to the phase of Γ_{12} being determined by $\lambda_t^2 = (V_{tb}^* V_{td})^2$, Equation (58), precisely the same same factor that determines if M_{12} from the dominant box contribution (cf. Equations (40) and (44)). Hence in this approximation one finds a_d = 0, i.e. CP invariance, which is easily understood as follows: While Γ_{12} and M_{12} are separately phase convention dependent, their ratio, as embodied in a_d is not. On the other hand, the approximation of assuming both the up and charm quark as essentially massless also amounts to assuming they are degenerate, and in this case we know that it is possible to redefine quark field phases so that the CKM matrix is real and the charged current weak interaction CP-invariant: Thus it is not surprising that one ends up with a vanishing same-sign dilepton asymmetry. In fact, one can estimate the effect of taking the phase space factors into account for Γ_{12} in the following way: the $c\bar{c}$ contribution will be suppressed as compared to the other three and is neglected. We can still set $m_u^2 = 0$ while the $(u\bar{c})$ and $(c\bar{u})$ contributions will involve a factor something like $\lambda_u \lambda_c (1 - m_c^2/m_b^2)$: given all of this, we expect Γ_{12} to be given by

$$\Gamma_{12} = \frac{G_F^2}{8\pi} f_B^2 M_B^3 \lambda_t^2 \left[1 + 0 \left(\frac{\lambda_u \lambda_c}{\lambda_t^2} \frac{m_c^2}{m_b^2}\right)\right] \qquad (66)$$

leading to

$$\left| \mathrm{Im} \frac{\Gamma_{12}}{M_{12}} \right|_{B_d} \simeq 0(\pi) \frac{m_c^2}{m_t^2} \left| \mathrm{Im} \frac{\lambda_u \lambda_c}{\lambda_t^2} \right| \qquad (67)$$

One sees that in magnitude (powers of λ in the Wolfenstein parameterization) $\lambda_u \lambda_c / \lambda_t^2 \simeq O(1)$ so the Standard Model expectation is very small

$$|a_d| \simeq O(\pi) \frac{m_c^2}{m_t^2} \lesssim O(\text{few} \times 10^{-3}) \qquad (68)$$

which should be reliable as to the order of magnitude. In fact, one can also give an estimated upper bound on the magnitude of a_d that could arise as a result of physics beyond the Standard Model with three generations: simply write first

$$\left|\text{Im}\, \Gamma_{12}/M_{12}\right| < \left|\Gamma_{12}/M_{12}\right|$$

while

$$\left|\frac{\Gamma_{12}}{M_{12}}\right| = \frac{|\Gamma_{12}|}{\Gamma} \cdot \frac{\Gamma}{|M_{12}|} = 2 \left|\frac{\Delta\Gamma}{2\Gamma}\right| \; \left|\Delta m/\Gamma\right|^{-1} \qquad (69)$$

Now note that only physical intermediate states contribute to $\Delta\Gamma$ so this quantity is unlikely to be affected by new physics, whatever this may be. Using our previous estimate of perhaps a percent or so for $|\Delta\Gamma/2\Gamma|$, and the known value of $(\Delta m/\Gamma) > 1/2$ it follows that no matter what

$$|a_d| \lesssim 5\ \%, \qquad (70)$$

say. This is still disappointingly small but could be a full order of magnitude above the three generation Standard Model expectation: it would clearly be of great significance to observe a dilepton asymmetry of the magnitude suggested in (70), but can this be done? It is elementary to establish that to observe an asymmetry "a" at the "n" - σ level requires a number $N_{B\bar{B}}$ of B-meson pairs such that

$$N_{B\bar{B}} \times (\text{branching ratios} \times \text{efficiencies} \times \ldots) \gtrsim n^2/a^2 \qquad (71)$$

For a (coherent odd-C) dilepton asymmetry a_d to be established at the 3σ level, with a mixing probability of 20% and detection efficiencies of 50%, say, would require

$$N_{B\bar{B}} \gtrsim 10^4/a_d^2 \tag{72}$$

With $a_d \sim 3\%$ (an order of magnitude greater than the Standard Model expectation), this means at least 10^7 $B\bar{B}$ pairs are required, which might be accessible with a year of running at a dedicated B-meson factory: reaching this level of sensitivity would be fiendishly difficult, but observation of a signal under these circumstances would be a clear indication of physics beyond the Standard Model (observing a_d at the level of the Standard Model expectations would require at least 10^9 $B\bar{B}$ pairs!).

To summarize what was discussed so far, the Standard Model expectation for CP-non invariance in B_d^o - $\bar{B}_d^o$ mixing is at least an order of magnitude smaller than its counterpart in K^o - $\bar{K}^o$ mixing as given in Equation (26), with that in B_s^o - $\bar{B}_s^o$ mixing even smaller still.

We now turn to an abbreviated discussion of an alternate set of CP-non invariant signals in B-meson physics, namely in B decay amplitudes (i.e. the analogs of ϵ' for kaons). It is well-known that while the CPT theorem implies the equality of lifetimes for particle and antiparticle, CP non-invariance can lead to differences in partial decay rates: quite generally, this requires the interference of two decay amplitudes. To illustrate this, consider the amplitude for decay of a B-meson (charged or neutral) into a particular final state "f" written as

$$A(B \to f) = A_1 e^{i\delta_1} + A_2 e^{i\delta_2} \tag{73}$$

The two interfering amplitudes may correspond to two different isospin channels, for example, and where the strong interaction final state phases have been factored out: for B-decays, these are unknown (and no way is known to exist to estimate them reliably: this will be a problem!). In practice, the two amplitudes can arise from two different single quark decay diagrams, or from a quark decay and a quark annihilation diagrams, or from a quark decay and a penguin diagram, etc. The analogous amplitude for $\bar{B} \to \bar{f}$ decay is written as,

$$A(\bar{B} \to \bar{f}) = \bar{A}_1 e^{i\delta_1} + \bar{A}_2 e^{i\delta_2} \tag{74}$$

where CPT invariance of the weak Hamiltonian together with strong interaction S-matrix unitarity and an allowed choice of overall phase imply

$$\bar{A}_1 = A_1^*, \ \bar{A}_2 = A_2^* \tag{75}$$

In fact, in general, a different choice of phase could be made to make one or the other of the amplitudes A_1 and A_2 purely real (this is what is usually done in the analysis leading to the usual definition of ϵ' in neutral kaon decays into two pions): we will not do so here. One readily discovers that

$$\Gamma(B \to f) - \Gamma(\bar{B} \to \bar{f}) \ \alpha \ (\text{Im}\, A_1 A_2^*) \sin(\delta_2 - \delta_1) \tag{76}$$

which can only be non-vanishing to the extent that the two amplitudes A_1 and A_2 are <u>not</u> relatively real and that the final state interaction phases are different. Case by case, the relative phase of the weak amplitudes A_1 and A_2 can be estimated by examination of the CKM angle factors appearing in the respective contributing diagrams at the quark level, but as mentioned before the final state interaction phases are unknown. Thus, no firm theoretical estimates can be said to exist for CP-non-invariant effects of this type.

Rather, to expose CP-non invariance in B-decay amplitudes consider the following situation: the decay of a neutral B-meson into a final state f that is a CP-eigenstate $|f^{CP}\rangle = \pm |f\rangle$ <u>and</u> which proceeds thought only one amplitude. Examples of processes satisfying these criteria are for example B^o, $\bar{B}^o \to J/\psi\, K_S\, \pi^+\pi^-$, $p\bar{p}$. We may then consider the time integrated asymmetry

$$\delta_f = \frac{\Gamma(B^o \to f) - \Gamma(\bar{B}^o \to f)}{\Gamma(B^o \to f) + \Gamma(\bar{B}^o \to f)} \tag{77}$$

of the partial rates of B-mesons produced as pure B^o and $\bar{B}^o$ states at t=0 respectively that can arise as a result of the interference of the mixing and decay amplitudes. Note that in this case because only one weak decay amplitude contributes and $|f^{CP}> = \pm|f>$, one can use the CPT theorem to establish

$$|A(B^o \to f)| = |A(\bar{B}^o \to f)| \tag{78}$$

and final state strong interaction phases are irrelevant. Denoting the respective amplitudes A and $\bar{A}$ for short, one has $|A/\bar{A}| = 1$ or $A = e^{i\xi}\bar{A}$ where ξ depends on the relative phase convention for $|B>$ and $|\bar{B}^o>$. Under these conditions, using Equations (14) and (18) in particular, it is easy to establish that

$$\delta_f = - \frac{x\text{Im}\,\eta_B\bar{A}/A}{1 + x^2} \tag{79}$$

where η_B is the quantity defined in Equation (8) for B-mesons and $x = (\Delta m/\Gamma)_B$ is the mixing parameter. Here, we have made the further very good approximation that $|\eta_B|^2 = 1$ (cf. Equations (24) and (68)) so that we take

$$\eta_B^{-1} = \eta_B^* \tag{80}$$

in deriving Equation (79) above. Note that both η_B and $\bar{A}/A$ are pure phase factors that are separately phase convention dependent, but their <u>product</u> is a physically significant phase factor that is <u>independent</u> of the choice of relative phase for $|B^o>$ and $|\bar{B}^o>$. Note also that $\delta_f \to 0$ for both limiting cases $x \to 0$ (no mixing) and $x \to \infty$ (complete mixing) so that while a sizeable mixing is necessary, there is such a thing as too much of a good thing here! In fact, the optimal value of x in δ_f is $|x| = 1$, quite close to the $B_d^o - \bar{B}_d^o$ value, while for the B_s^o - $\bar{B}_s^o$ system $|x| > 2$ making it less suitable for the study of this time integrated asymmetry.

The measurement of δ_f requires the tagging of the initially produced meson as a B^o or $\bar{B}^o$ by the identification of the $\bar{B}$ or B partner particle produced with it at t = 0, (this can be charged or neutral) for example for means of a semileptonic or other readily identifiable decay: if this is done starting with a coherently produced B^o $\bar{B}^o$ pair (e.g. in a C-odd state from $\Upsilon(4s)$ decay), the expression for δ_f given above no longer holds. For example it is easy to see that with leptonic tagging, for an initial C-odd coherent superposition one finds

$$\delta_f(\text{C-odd; coherent}) = \frac{\Gamma(B_d^o \to f;\ \ell^-) - \Gamma(\bar{B}_d^o \to f;\ \ell^+)}{\Gamma(B_d^o \to f;\ \ell^-) + \Gamma(\bar{B}_d^o \to f;\ \ell^+)} \tag{81}$$

$$= \frac{1 - |\eta_B|^4}{1 + |\eta_B|^4} = a_d \ !$$

i.e. precisely the same result as the corresponding dilepton asymmetry and of a completely different nature as compared to the result embodied in Equations (77) and (79) appropriate to initial states unhindered by quantum mechanical correlations! We leave it as an exercise to the reader to evaluate the time-integrated asymmetries a_d and δ_f as well as the like-sign relative to the unlike-sign dilepton yields when the initial state is a B_d^o $\bar{B}_d^o$ pair produced in a C-even eigenstate, as could be achieved, for example, by running on the $\Upsilon(5s)$ resonance and considering the decays $\Upsilon(5s) \to B_d^o \bar{B}_d^{o*} + B_d^{o*} \bar{B}_d^o \to B_d^o \bar{B}_d^o \gamma$ with photon tagging which guarantees the B_d^o $\bar{B}_d^o$ pair to be in the C-even state. He or she will then discover how the lack of the EPR correlation present in the C-odd case changes the situation and hopefully marvel at the results! Incidentally, running at the $\Upsilon(5s)$ may be a good start for a search for B_s^o - $\bar{B}_s^o$ mixing!

We now return to the single neutral B meson asymmetry δ_f of Equation (79) as applicable to a B_d^o or $\bar{B}_d^o$ meson in an initially incoherently produced $B\bar{B}$ pair, and give an estimate of its magnitude. For $f = J/\psi\ K_S$, the single weak decay amplitude corresponds to the b quark decay $b \to c + W^* \to c + \bar{c}s$ for $\bar{B}^o_d$ and $\bar{b} \to \bar{c} + W^* \to \bar{c} + \bar{s}c$ for B^o_d: Thus, one finds

$$(\bar{A}/A)_{J/\psi K_S} = V_{cb}V^*_{cs}/V^*_{cb}V_{cs} \tag{82}$$

while for f either $\pi^+\pi^-$ or $p\bar{p}$ the single amplitude corresponds to $b \to u\,\bar{u}d$ and $\bar{b} \to \bar{u}\,u\bar{d}$ for $\bar{B}^o_d$ and B^o_d respectively, which leads to

$$(\bar{A}/A)_{\pi^+\pi^-,p\bar{p}} = V_{ub}V^*_{ud}/V^*_{ub}V_{ud} \tag{83}$$

Correspondingly, with

$$\eta_B = (\mathcal{M}_{21} / \mathcal{M}_{12})^{1/2} \simeq (M^*_{12}/M_{12})^{1/2} \tag{84}$$

one has

$$\eta_B = V_{td}V^*_{tb}/V^*_{td}V_{tb} \tag{85}$$

The reader may wish to verify that in all cases the quantity Im $\eta_B\ \bar{A}/A$ is in fact invariant under phase redefinitions of any and all up-type and down-type quarks separately (this is not obvious in the case of $J/\psi K_S$!).

The upshot of all of this is that δ_f of this type is calculable from a knowledge of the CKM elements, and asymmetries as large as the order of 10% are possible. There is however a price to pay in determining how many B meson pairs are required to measure such an asymmetry, namely, small branching ratios and less than perfect tagging efficiencies, as per Equation (71): consider the $J/\psi\ K_S$ case for purposes of illustration. Assume that $J/\psi\ K_S$ will be detected using its decays into the final state $\ell^+\ell^-\ \pi^+\pi^-$ with $\ell = e, \mu$. Assuming a branching ratio of about 10^{-3} for $B^o_d \to J/\psi\ K_S$ (that is, the same as the measured value for $B^-_u \to J/\psi\ K^-$) and with the other known branching ratios gives an effective overall branching ratio of 10^{-4}: further assume a 50% detection efficiency as well as a 10% tagging efficiency for the initial B or $\bar{B}$ partner (recall the semileptonic branching ratio into e and μ is about 20%). Put all this together with the requirement of a 3σ effect and obtain

$$N_{B\bar{B}} \gtrsim \frac{2 \times 10^6}{\delta_f^2} \tag{86}$$

with δ_f = 30%, say, this means that over 10^7 pairs are required to see this particular instance of CP non-invariance at the level predicted by the Standard Model. Numbers of this magnitude recur over and over again, and point to the necessity of a B-factory.

We have limited our discussion of mixing and of decay asymmetries to time-integrated quantities: one can hope that the development of even more accurate methods of vertex detection will eventually allow the observation of the time dependence of B-$\bar{B}$ oscillations and CP-non-invariant asymmetries, much as is routinely done in the K^o - $\bar{K}^o$ system.

5. CONCLUSIONS AND ACKNOWLEDGEMENTS

The discovery of B_d^o-$\bar{B}_d^o$ mixing by the ARGUS collaboration has opened a new door for the investigation of the physics of the Standard Model, which has so far passed all tests with flying colours. It also points to a new venue for the search for physics beyond the Standard Model, by adding rare B-meson processes to the rare K-meson decays which comprise part of our arsenal of probes of new short distance phenomena: it seems clear that a dedicated B-factory will be an ideal (and relatively inexpensive) tool for significant research in experimental high energy physics in the coming decade. Unless Canada can secure a significant role in the siting and running of the SSC, the construction of a B-factory should in my opinion be the major goal of Canadian efforts in experimental high energy physics in the nineties.

This work was supported in part by the United States Department of Energy under grant DE-AC02-83ER-40105 as well as by a Presidential Young Investigator Award.

SELECTED REFERENCES

In keeping with the informal tone of these notes, the following comprises a decidedly incomplete list of references. A more complete bibliography will be found in a forthcoming review article "Weak Mixing and the Violation of CP-Invariance in B-Meson Physics", to be published in Surveys in High Energy Physics.

On the quantum mechanics of the mixing of neutral pseudoscalar mesons:

Gell-Mann, M. and Pais, A. Phys. Rev. 97, 1387(1955)

Day, T.B., Phys. Rev. 121, 1204(1961)

Inglis, D.R., Rev. Mod. Phys. 33, 1(1961)

Six, J., Phys. Lett. B114, 200(1982)

Datta, A. and Home, D., Phys. Lett. A119, 3(1986)

Squires, E. and Siegwart, D., Phys. Lett. A126, 73(1987)

Finkelstein, J. and Stapp, H.P., Phys. Lett. A126, 159(1987)

A complete, up-to-date survey of experimental results on CP-non invariance in the neutral kaon system:

Steinberger, J., "Experimental Status of CP-violation", Invited Talk given at the Rencontre de Physique de la Vallee d'Aoste, La Thiule, Italy, March 1988, CERN-EP/88-66, to be published in the proceedings.

Early work on dilepton signals of weak mixing

Okun, L.B., Zakharov, V.I. and Pontecorvo, B.M., Lett. Nuovo. Cim. 13, 218(1975)

Pais, A. and Treiman, S.B., Phys. Rev. D12, 2744(1975)

The suggestion that observation of a like sign dilepton signal for B^o-$\bar{B}^o$ mixing could provide a measurement m_t was first made in

Ellis, J., Gaillard, M.K., Nanopoulos, D.V. and Rudaz, S., Nucl. Phys. B131, 285(1977)

The expression "Penguin Diagram" also made its first appearance in that paper: after ten years, this was followed by

Ellis, J., Hagelin, J.S. and Rudaz, S., Phys. Lett. B192, 201(1987)

prompted by the ARGUS discovery of B_d^o - $\bar{B}_d^o$ mixing,

Albrecht, H. *et al*., Phys. Lett. B192, 245(1987)

A more detailed analysis of the implications of B_d^o - $\bar{B}_d^o$ mixing and the measurement of ϵ'/ϵ_o is contained in

Ellis, J., Hagelin, J.S., Rudaz, S. and D.D. Wu, Nucl. Phys. B304, 205(1988)

Where extensive references can be found. Some other papers concerned with the same topics, in alphabetical order,

Altarelli, G. and Franzini, P., Z. Phys. C37, 271(1988)

Barger, V., Han, T., Nanopoulos, D.V. and Phillips, R.J.N., Phys. Lett. 194B, 312(1987)

Bigi, I.I. and Sanda, A., Phys. Lett. B194, 307(1987)

Chau, L.L. and Keung, W.Y., Phys. Rev. Lett. 59, 958(1987)

Du, D.S. and Zhao, Z.-Y., Phys. Rev. Lett. 59, 1972(1987)

Khoze, V.A. and Uraltsev, N.G. Leningrad preprint 1290(1987)

and many others.

The topic of CP-non-invariance in B-physics has been a special preoccupation of

Bigi, I.I. and Sanda, A.I., Nucl. Phys. B193, 85(1981); Phys. Rev. D29, 1393(1984)

and Chau, L.L. and Cheng, H.Y., Phys. Rev. Lett. 53, 1037(1984); Phys. Lett. B165, 429(1985).

Complete review of the CP-non invariant phenomena:

Donoghue, J.F., Holstein, B. and Valencia, G., Int. J. Mod. Phys. A2, 319(1987);

Bigi, I.I., Khoze, V.A., Uraltsev, N.G. and Sanda, A.I., SLAC-PUB-4476(1987), to be published in "CP-violation", C. Jarlskog ed. (World Scientific, Singapore, 1988)

and also

Rosner, J.L., Sanda, A.I. and Schmidt, M.P., Chicago preprint EFI-88-12, in the Proceeding of High Sensitivity Beauty (?) Physics at Fermilab, to be published in the Proceedings.